NMR and Chemistry

NMR and Chemistry

An introduction to
modern NMR spectroscopy

Third edition

J. W. Akitt

CHAPMAN & HALL
London · Glasgow · New York · Tokyo · Melbourne · Madras

Published by Chapman & Hall, 2–6 Boundary Row, London SE1 8HN

Chapman & Hall, 2–6 Boundary Row, London SE1 8HN, UK

Blackie Academic & Professional, Wester Cleddens Road, Bishopbriggs, Glasgow, G64 2NZ, UK

Chapman & Hall, 29 West 35th Street, New York NY 10001, USA

Chapman & Hall Japan, Thomson Publishing Japan, Hirakawacho Nemoto Building, 7F, 1-7-11 Hirakawa-cho, Chiyoda-ku, Tokyo 102, Japan

Chapman & Hall Australia, Thomas Nelson Australia, 102 Dodds Street, South Melbourne, Victoria 3205, Australia

Chapman & Hall India, R. Seshadri, 32 Second Main Road, CIT East, Madras 600 035, India

First edition 1973

Second edition 1983

Third edition 1992

© 1973, 1983, 1992   J. W. Akitt

Typeset in 10/12 Times by Thomson Press (India) Ltd, New Delhi
Printed in Great Britain by T. J. Press (Padstow) Ltd

ISBN 0 412 37260 6

A catalogue record for this book is available from the British Library

Library of Congress Cataloging-in-Publication Data
Akitt, J. W.
    NMR and chemistry: an introduction to modern NMR spectroscopy/
J. W. Akitt.—3rd ed.
        p.   cm.
    Includes bibliographical references and index.
    1. Nuclear magnetic resonance spectroscopy.   I. Title.
QD96.N8A37 1992   92–3527
543'.0877—dc20   CIP

Contents

Preface to the third edition

It is a little over nine years since I wrote the preface to the second edition of this book and the story is still that the NMR community continues in its inventive and innovative manner. Certainly, the chapter headed 'Some new and exciting techniques in NMR' has been replaced by a series of separate chapters giving much more detailed coverage of two-dimensional NMR, of imaging and of high-resolution solid-state NMR. These techniques are no longer new but remain definitely exciting. Two-dimensional techniques are enabling much more complex spectra to be tackled and interpreted, and experimental variations giving access to new types of information are appearing almost daily. Imaging is important in medicine but has now moved into technological and research areas. Solid-state NMR has become routine and ^{13}C spectra are being published which to a casual glance look quite like their liquid-state counterparts. Multinuclear NMR, which not so long ago meant effectively ^{1}H and ^{13}C NMR, now encompasses the whole range of accessible nuclei and appears without special comment in the scientific journals. The potential for advance remains immense and there seem to be few areas where an NMR technique cannot be used with some chance of success.

All this has proved a problem for the author in deciding what to leave in and what new material to include. In an introductory text one has to start in a very straightforward manner, the object being to familiarize newcomers to the subject with its possibilities and attempt to provide explanations of the complexities and theories needed to progress. In addition one has to describe the current state of things, and indicate what people are actually doing, so that if a student happens to open the page of a journal he or she will have some inkling of what is being described. In bread-and-butter terms this means starting with the ^{1}H spectrum of, say, chloroethane (ethyl chloride), and working up to the solution of the spectra of amygdalin using perhaps five different two-dimensional methods – and this is quite a simple example when compared with what is being done in the field of the NMR of enzymes. I have thus omitted the applications chapter in this edition since the text is naturally multinuclear. The examples now are used to illustrate the chapters describing the various NMR parameters. There are two other major differences also: spin–spin coupling is treated more briefly since the detailed understanding of second-order effects is less necessary now that very-high-field spectrometers are more generally available. On the other hand, the chapter on dynamic systems has been much expanded to give this important area a coverage more adapted to that it deserves.

I am indebted to colleagues for comments and assistance: old colleagues at the University of Newcastle-upon-Tyne for Figs 2.13 and 3.12; Dr O. W. Howarth of Warwick University for Fig. 7.21 and for comments on Chapter 6; Dr B. E. Mann of Sheffield University for Fig. 2.9; Dr A. Römer of Cologne University for Examples 15 and 16; and Professor B. L. Shaw for Figs 3.13 and 6.18. I also give grateful acknowledgements to Varian Associates Ltd for permission to reproduce the spectra in Figs 3.4, 3.6, 3.18, the spectrum of 1-chloropropionic acid, and Examples 1 to 12; and to Bruker Spectrospin Ltd for permission to reproduce diagrams and spectra in their application notes and reports in Figs 7.2, 7.3, 7.19, 9.5, 9.7, 10.2, 10.3, 10.4 and 10.9 and to Professor A. E. Merbach of the University of Lausanne for assistance with locating a figure.

J.W.A.
Séez
1992

Preface to the second edition

It is just ten years since I wrote the preface to the first edition of this book. The intervening decade has, however, seen such an explosive development of the subject that it has changed almost out of all recognition. Certainly some material has had to be completely replaced by new and the text has had to be extensively rewritten to accommodate current concepts. The first edition contained a mention of Fourier transform techniques, and of superconducting magnets, and these two fields have both developed extremely rapidly because it was realized that they would make possible some real advances in chemical research. The new ^{13}C spectroscopy of organic molecules on the one hand increased the scope of the technique for structural determination, and high field proton spectroscopy on the other enabled the problem of the solution structure of large, biologically important molecules to be tackled. It was also increasingly realized that the Fourier transform pulse techniques allowed precise manipulation of nuclear spins so that many different new relaxation or double resonance experiments became possible. Such advances have also proved to be informative when applied to the less popular nuclei and so multinuclear flexibility was introduced into the new, powerful NMR spectrometers, which can, in principle, carry out most likely experiments with every magnetically active nucleus in the periodic table.

Development has, however, not stopped here. The difficult field of the high resolution study of solid samples is being successfully ploughed; the separation of shift and coupling parameters in a two-dimensional experiment is now possible and the biologists have started to look at somewhat unusual samples such as anaesthetized live rats.

One particularly interesting development which is being vigorously pursued is that of whole body imaging. It has proved possible to map out the proton density of the water in the body using pulsed NMR techniques in conjunction with specially contoured magnetic fields. This gives a thin cross-section of selected parts of human subjects which can be obtained quickly and safely and gives information which is likely to be complementary to that obtained by X-rays. Thus our technique has expanded and moved into an area which can be seen to be vital to the whole of mankind. Practising spectroscopists will of course be able to point to many areas where NMR has proved invaluable to science and industry, and so to mankind in general. Such advances are unfortunately not immediately obvious to the public at large and we should welcome the advent of whole body imaging, as a visible testament to the utility of pure science. The physicists who first detected nuclear magnetic resonances are unlikely to have had concern for anything but the details their investigations revealed about the nucleus of the atom. When it became apparent to chemists that the technique was capable of giving uniquely valuable information about the structure of molecules, then their demands led to the development of a spectrometer industry serving academic and industrial science; their quest for higher sensitivity led to the Fourier transform technique and this now is set to lead us into a new medical based industry. Such developments should be well pondered by those who would constrain pure scientific research.

I have left the general plan of the book much as in the first edition; theory, followed by examples of the use of the technique which serve also to reinforce the theory. The technique is approached as primarily a pulse spectrometry but it has proved necessary to retain mention of the old continuous wave methods. Continuous wave equipment is still with us and even the most advanced pulse spectrometer makes use of a continuous wave technique to adjust the magnetic field homogeneity and to provide a field-frequency lock.

I am again indebted to many colleagues for assistance and comments; Dr R. J. Bushby for Fig. 9.17, Professor N. N. Greenwood, Dr O. W. Howarth of Warwick University for Fig. 9.22, Dr J. D. Kennedy for Figs 9.37–9.39 and 9.53, Dr B. E. Mann of Sheffield University for Fig. 9.32, Professeur G. J. Martin of the Université de Nantes for supplying the spectra to illustrate Fig. 4.20, Dr A. Römer of Cologne University for Figs 9.14–9.16 and Professor B. L. Shaw for Figs 9.25, 9.44–9.46. I am doubly indebted to Dr Kennedy for reading the entire manuscript and for suggesting some valuable additions. I also give grateful acknowledgement to Varian International AG for permission to reproduce the spectra in Figs 3.3, 3.5, 3.14, 3.17, 9.2–9.10, 9.13; to Bruker-Spectrospin Ltd for permission to reproduce diagrams and spectra in their application notes as Figs 7.4, 7.5, 8.9–8.16; to Heyden and Son Ltd (John Wiley and Sons, Inc.) for permission to reproduce Figs 4.4, 4.17, 4.20, 8.4, 8.5 and Table 7.1; to the American Chemical Society for permission to reproduce Figs 3.22, 4.19, 7.6, 8.3, 9.20, 9.21, 9.27, 9.28, 9.34, 9.35, 9.42, 9.43, 9.48 and 9.50; to Academic Press for permission to reproduce Figs 5.9, 5.10, 7.10, 7.15, 7.16, 7.17 and 9.41; to the American Association for the Advancement of Science for permission to reproduce Figs 8.6 and 8.7; to the Royal Society of Chemistry for permission to reproduce Figs 3.11, 7.2, 8.2, 8.8 and 9.23; to the American Institute of Physics for permission to reproduce Figs 6.2, 8.17, 9.29, 9.30, 9.31, 9.36 and 9.51; to Elsevier Sequoia SA for permission to reproduce Fig. 9.26; and the Royal Society for permission to reproduce Fig. 3.21.

J.W.A.
Leeds
March 1982

Preface to the first edition

About 20 years have elapsed since chemists started to take an interest in nuclear magnetic resonance spectroscopy. In the intervening period it has proved to be a very powerful and informative branch of spectroscopy, so much so that today most research groups have access to one or more spectrometers and the practising chemist can expect constantly to encounter references to the technique. There is a considerable number of textbooks available on the subject but these are invariably written primarily either for the specialist or for the graduate student who is starting to use the technique in his research. The author has, however, always felt that a place existed for a non-specialist text written for the undergraduate student giving an introduction to the subject which embraced the whole NMR scene and which would serve as a basis for later specialization in any of the three main branches of chemistry.

With this in mind the book has been written in two sections. The first covers the theory using a straightforward non-mathematical approach which nevertheless introduces some of the most modern descriptions of the various phenomena. The text is illustrated by specific examples where necessary. The second section is devoted to showing how the technique is used and gives some more complex examples illustrating for instance its use for structure determination and for measurements of reaction rates and mechanisms. A few problems have been included but the main purpose of the book is to demonstrate the many and varied present uses of NMR rather than to teach the student how to analyse a spectrum in detail. This is done best if it is done concurrently with a student's own research.

I am indebted to Dr K. D. Crosbie, Professor N. N. Greenwood, Dr B. E. Mann, and to Professor D. H. Whiffen who read and criticized the manuscript and to many former colleagues at Newcastle-upon-Tyne for encouragement and for some of the examples used in the text. I also give grateful acknowledgement to Varian Associates Ltd for permission to reproduce the spectra in Figs 15, 17, 26, 27, 66 and 75 and to Bruker-Spectrospin Ltd for permission to reproduce the spectra in Figs 49, 50, 63, 64 and 75.

J.W.A.
Leeds,
January, 1972

1 The theory of nuclear magnetization

1.1   The properties of the nucleus of an atom

The chemist normally thinks of the atomic nucleus as possessing only mass and charge and is concerned more with the interactions of the electrons that surround the nucleus, neutralize its charge and give rise to the chemical properties of the atom. Nuclei, however, possess several other properties that are of importance to chemistry, and to understand how we use them it is necessary to know something more about them.

Nuclei of certain natural isotopes of the majority of the elements possess intrinsic angular momentum or spin, of total magnitude $\hbar[I(I + 1)]^{1/2}$. The largest measurable component of this angular moment is $I\hbar$, where I is the nuclear spin quantum number and $\hbar$ is the reduced Planck's constant, $h/2\pi$. I may have integral or half-integral values $(0, 1/2, 1, 3/2,...)$, the actual value depending upon the isotope. Since I is quantized, several discrete values of angular momentum may be observable and their magnitudes are given by $\hbar m$, where the quantum number m can take the values $I, I - 1, I - 2,..., -I$. There are thus $2I + 1$ equally spaced spin states of a nucleus with angular momentum quantum number I.

A nucleus with spin also has an associated magnetic moment μ. We define the components of μ associated with the different spin states as $m\mu/I$, so that μ also has $2I + 1$ components. In the absence of an external magnetic field the spin states all possess the same potential energy, but take different values if a field is applied. The origin of the NMR technique lies in these energy differences, though we must defer further discussion of this until we have defined some other basic nuclear properties.

The magnetic moment and angular momentum behave as if they were parallel or antiparallel vectors. It is convenient to define a ratio between them which is called the magnetogyric ratio, γ:

$$\gamma = \frac{2\pi}{h}\frac{\mu}{I} = \frac{\mu}{I\hbar} \tag{1.1}$$

γ has a characteristic value for each magnetically active nucleus and is positive for parallel and negative for antiparallel vectors. We will see that the sign of γ

influences both spin–spin coupling and the way energy is exchanged between spins.

If $I > 1/2$ the nucleus possesses in addition an electric quadrupole moment, Q. This means that the distribution of charge in the nucleus is non-spherical and that it can interact with electric field gradients arising from the electric charge distribution in the molecule. This interaction provides a means by which the nucleus can exchange energy with the molecule in which it is situated and affects certain NMR spectra profoundly.

Some nuclei have $I = 0$. Important examples are the major isotopes ^{12}C and ^{16}O, which are both magnetically inactive – a fact that leads to considerable simplification of the spectra of organic molecules. Such nuclei are, of course, free to rotate in the classical sense, but this must not be confused with the concept of quantum-mechanical 'spin'. The nucleons, that is the particles such as neutrons and protons which make up the nucleus, possess intrinsic spin in the same way as do electrons in atoms. Nucleons of opposite spin can pair, just as do electrons, though they can only pair with nucleons of the same kind. Thus in a nucleus with even numbers of both protons and neutrons all the spins are paired and $I = 0$. If there are odd numbers of either or of both, then the spin is non-zero, though its actual value depends upon orbital-type inter-nucleon interactions. Thus we build up a picture of the nucleus in which the different resolved angular momenta in a magnetic field imply different nucleon arrangements within the nucleus, the number of spin states depending upon the number of possible arrangements. If we add to this picture the concept that s bonding electrons have finite charge density within the nucleus and become partly nucleon in character, then we can see that these spin states might be perturbed by the hybridization of the bonding electrons and that information derived from the nuclear states might lead indirectly to information about the electronic system and its chemistry.

The most important nuclear properties of the elements are listed in Table 1.1. This gives the atomic weight of the nuclear isotope of the element listed, and where there is more than one magnetically active isotope this is indicated by a weight in parentheses. The isotope listed is the one most usually used, though the unlisted ones are in some cases equally usable. The next column gives the spin quantum number, I, followed by the natural abundance of the isotope. The receptivity or natural signal strength of the nucleus is given relative to that of ^{13}C, which itself gives a fairly weak signal, and this figure is made up of the intrinsic sensitivity of the nucleus (high if the magnetic moment is high) weighted by its natural abundance. Some elements are used in enriched forms (particularly ^{2}H and ^{17}O), when the receptivity is, of course, substantially higher. The data are completed by the quadrupole moment (where $I > 1/2$) and the resonant frequency in a particular magnetic field. This can be determined to a much higher precision than shown, for a resonance in an individual compound, and it should be remembered that each nucleus will have a range of frequencies due to the chemical shift effect. The frequency is proportional to the magnetogyric ratio.

Table 1.1 Nuclear properties of some of the elements

Element	Atomic weight	Spin, I	Natural abundance (%)	Receptivity (^{13}C = 1.00)	Quadrupole moment (10^{-30} m^2)	Resonant frequency (MHz) at 2.348 T
Hydrogen	1	1/2	99.985	5670	None	100.00
Deuterium	2	1	0.015	0.0082	0.287	15.35
Tritium	3	1/2	Radioactive	—	None	106.66
Helium	3	1/2	0.00014	0.0035	None	76.18
Lithium	7(6)	3/2	92.58	1540	−3.7	38.87
Beryllium	9	3/2	100	78.8	5.3	15.06
Boron	11(10)	3/2	80.42	754	4.1	32.08
Carbon	13	1/2	1.108	1.00	None	25.15
Nitrogen	14	1	99.63	5.70	1.67	7.23
Nitrogen	15	1/2	0.37	0.022	None	10.14
Oxygen	17	5/2	0.037	0.061	−2.6	13.56
Fluorine	19	1/2	100	4730	None	94.09
Neon	21	3/2	0.257	0.0036	9	7.90
Sodium	23	3/2	100	524	10	26.43
Magnesium	25	5/2	10.13	1.54	22	6.13
Aluminium	27	5/2	100	1170	14	26.08
Silicon	29	1/2	4.7	2.1	None	19.87
Phosphorus	31	1/2	100	377	None	40.48
Sulphur	33	3/2	0.76	0.098	−6.4	7.67
Chlorine	35(37)	3/2	75.53	20.2	−8.2	9.81
Potassium	39	3/2	93.1	2.69	5.5	4.67
Calcium	43	7/2	0.145	0.053	−5	6.74
Scandium	45	7/2	100	1720	−22	24.33
Titanium	49(47)	7/2	5.51	1.18	24	5.64
Vanadium	51(50)	7/2	99.76	2170	−5.2	26.35
Chromium	53	3/2	9.55	0.49	−15	5.64
Manganese	55	5/2	100	1014	40	24.84
Iron	57	1/2	2.19	0.00425	None	3.24

Continued

Table 1.1 (Contd.)

Element	Atomic weight	Spin, I	Natural abundance (%)	Receptivity ($^{13}C = 1.00$)	Quadrupole moment (10^{-30}m^2)	Resonant frequency (MHz) at 2.348 T
Cobalt	59	7/2	100	1560	42	23.73
Nickel	61	3/2	1.19	0.24	16	8.93
Copper	63(65)	3/2	69.09	368	−22	26.51
Zinc	67	5/2	4.11	0.67	15	6.25
Gallium	71(69)	3/2	39.6	322	11	30.58
Germanium	73	9/2	7.76	0.62	−17	3.48
Arsenic	75	3/2	100	144	29	17.18
Selenium	77	1/2	7.58	3.02	None	19.07
Bromine	81(79)	3/2	49.46	279	27	27.10
Krypton	83	9/2	11.55	1.24	27	3.86
Rubidium	87(85)	3/2	27.85	280	13	32.84
Strontium	87	9/2	7.02	1.08	16	4.35
Yttrium	89	1/2	100	0.676	None	4.92
Zirconium	91	5/2	11.23	6.05	−21	9.34
Niobium	93	9/2	100	2770	−32	24.55
Molybdenum	95(97)	5/2	15.72	2.92	−1.5	6.55
Ruthenium	99(101)	5/2	12.72	0.815	7.6	4.61
Rhodium	103	1/2	100	0.18	None	3.16
Palladium	105	5/2	22.23	1.43	65	4.58
Silver	109(107)	1/2	48.18	0.28	None	4.65
Cadmium	113(111)	1/2	12.26	7.69	None	22.18
Indium	115(113)	9/2	95.72	1920	86	22.04
Tin	119(115,117)	1/2	8.58	25.7	None	37.29
Antimony	121(123)	5/2	57.25	530	−33	24.09
Tellurium	125(123)	1/2	6.99	12.8	None	31.55
Iodine	127	5/2	100	541	−79	20.15
Xenon	129(131)	1/2	26.44	32.4	None	27.86
Caesium	133	7/2	100	275	−0.3	13.21

Barium	137(135)	3/2	11.32	4.5	28	11.19
Lanthanum	139(138)	7/2	99.91	343	22	14.24
Praseodymium	141	5/2	100	1620	−4.1	29.03
Neodymium	145(143)	7/2	8.3	0.393	−25	3.41
Samarium	149(147)	7/2	13.83	0.665	5.6	3.43
Europium	151(153)	5/2	47.82	464	114	24.48
Gadolinium	155(157)	3/2	14.73	0.124	160	3.09
Terbium	159	3/2	100	394	134	24.04
Dysprosium	163(161)	5/2	24.97	1.79	251	4.77
Holmium	165	7/2	100	1160	349	21.34
Erbium	167	7/2	22.94	0.665	283	2.90
Thulium	169	1/2	100	2.89	None	7.99
Ytterbium	171(173)	1/2	14.31	4.5	None	17.70
Lutetium	175(176)	7/2	97.41	173	346	11.43
Hafnium	177(179)	7/2	18.50	1.47	330	4.06
Tantalum	181	7/2	99.988	213	330	12.13
Tungsten	183	1/2	14.28	0.0608	None	4.22
Rhenium	187(185)	5/2	62.93	511	220	23.05
Osmium	187(189)	1/2	1.64	0.00115	None	2.28
Iridium	193(191)	3/2	62.7	0.122	78	1.90
Platinum	195	1/2	33.8	19.9	None	21.50
Gold	197	3/2	100	0.153	55	1.75
Mercury	199(201)	1/2	16.84	5.68	None	17.87
Thallium	205(203)	1/2	70.5	807	None	57.63
Lead	207	1/2	22.6	11.9	None	20.92
Bismuth	209	9/2	100	819	−37	16.36
Uranium	235	7/2	0.72	0.0054	455	1.84

Source: Mason (1987) *Multinuclear NMR*, Plenum Press, New York.

The first point to note about this list is that almost all the elements are represented, the only missing ones being argon, technetium, cerium and promethium, and that, in principle, virtually the whole of the periodic table can be studied by NMR. Indeed, with modern instrumentation, this is now realizable, though there are some cases, notably nuclei with very high quadrupole moments, where study in the liquid state is not rewarding. The usefulness of a nucleus to the NMR spectroscopist depends in the first place upon the chemical importance of the atom it characterizes and then upon its receptivity. Thus the extreme importance of carbon spectroscopy for understanding the structures of organic molecules has led to technical developments that have overcome the disadvantages of its poor receptivity, so that ^{13}C NMR is now commonplace. Hydrogen, with its very high receptivity, has, of course, been studied right from the emergence of NMR spectroscopy as a technique useful to chemists, and proton spectroscopy, as it is often called (proton $= {}^1H$ – the term is commonly used by NMR spectroscopists when discussing the nucleus of neutral hydrogen), has been used for the identification of the majority of organic compounds and many inorganic ones, and for the physical study of diverse systems. Other much-studied nuclei are ^{11}B in the boron hydrides or carboranes, ^{19}F in the vast array of fluoro organics and inorganics and ^{31}P in its many inorganic and biochemical guises. However, even quite low receptivity is no longer an insurmountable obstacle and, if a chemical problem is presented that is capable of solution by NMR spectroscopy, then whatever nucleus may require to be observed, the appropriate effort may well bring the desired rewards.

1.2   The nucleus in a magnetic field

If we place a nucleus in a magnetic field B_0 it can take up $2I + 1$ orientations in the field, each one at a particular angle θ to the field direction and associated with a different potential energy. The energy of a nucleus of magnetic moment μ in field B_0 is $-\mu_z B_0$, where μ_z is the component of μ in the field direction. The energy of the various spin states is then

$$-\frac{m\mu}{I}B_0 \quad \text{or individually} \quad -\mu B_0, -\frac{I-1}{I}\mu B_0, -\frac{I-2}{I}\mu B_0, \text{ etc.}$$

The energy separation between the levels is constant and equals $\mu B_0/I$. This is shown diagrammatically in Fig. 1.1 for a nucleus with $I = 1$ and positive magnetogyric ratio. The value of m changes sign as it is altered from I to $-I$ and accordingly the contribution of the magnetic moment to total nuclear energy can be either positive or negative, the energy being increased when m is positive. The energy is decreased if the nuclear magnetic vectors have a component aligned with the applied field in the classical sense. An increase in energy corresponds to aligning the vectors in opposition to the field. Quantum mechanics thus predicts a non-classical situation, which can only arise because

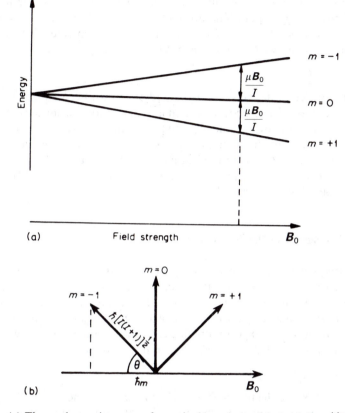

Fig. 1.1 (a) The nuclear spin energy for a single nucleus with $I = 1$ (e.g. ^{14}N) plotted as a function of magnetic field B_0. The two degenerate transitions are shown for a particular value of B_0. (b) The alignment of the nuclear vectors relative to B_0 that correspond to each value of m. The vector length is $\hbar[I(I + 1)]^{1/2}$ and its z component is $\hbar m$, whence $\cos \theta = m/[I(I + 1)]^{1/2}$.

of the existence of discrete energy states with the high-energy states indefinitely stable.

In common with other spectral phenomena, the presence of a series of states of differing energy in an atomic system provides a situation where interaction can take place with electromagnetic radiation of the correct frequency and cause transitions between the energy states. The frequency is obtained from the Bohr relation, namely

$$hv = \Delta E$$

where ΔE is the energy separation. For NMR

$$hv = \mu B_0/I$$

In this case the transition for any nuclear isotope occurs at a single frequency since all the energy separations are equal and transitions are only allowed between adjacent levels (i.e. the selection rule $\Delta m = \pm 1$ operates). The frequency relation is normally written in terms of the magnetogyric ratio (1.1), giving

$$v = \gamma B_0/2\pi \tag{1.2}$$

Thus the nucleus can interact with radiation whose frequency depends only on the applied magnetic field and the nature of the nucleus. Magnetic resonance spectroscopy is unique in that we can choose our spectrometer frequency at will, though within the limitation of available magnetic fields. The values of γ are such that for practical magnets the frequency for nuclei lies in the radio range between a present maximum of 600 megahertz (MHz) and a minimum of a few kilohertz (kHz).

1.3   The source of the NMR signal

The low frequency of nuclear magnetic resonance absorption indicates that the energy separation of the spin states is quite small. Since the nuclei in each of the states are in equilibrium, this suggests that the numbers in the different spin states will be similar, though if there is a Boltzmann distribution among the spin states then we can expect more nuclei to reside in the lowest energy states. For a system of spin-1/2 nuclei a Boltzmann distribution would give

$$N_u/N_l = \exp(-\Delta E/kT)$$

where N_u and N_l are the numbers of nuclei in the upper and lower energy states respectively, ΔE is the energy separation, k is the Boltzmann constant, and T is the absolute temperature. ΔE is given above as $\mu B_0/I$, which, for $I = 1/2$, equals $2\mu B_0$. Thus

$$N_u/N_l = \exp(-2\mu B_0/kT)$$

which since $N_u \approx N_l$ can be simplified to

$$N_u/N_l = 1 - 2\mu B_0/kT \tag{1.3}$$

For hydrogen nuclei in a magnetic field of 9.39 T (tesla), where the resonance frequency is 400 MHz, and at a temperature of 300 K, the quantity $2\mu B_0/kT$ has a value of about 6×10^{-5}, which means that the excess population in the lower energy state is one nucleus per 300 000. This is extremely small. The z components of the magnetic moments of the nuclei in the upper energy state are all cancelled by those of the lower energy state, only the small number excess in the lower energy state being able to give rise to an observable magnetic effect. This is a weak nuclear paramagnetism, which is only observable at low temperatures where the population difference is a maximum. For this reason we have to resort to a resonance technique in order to observe a signal.

If $I > 1/2$ one obtains a similar though more complex picture since the excess low-energy nuclei do not all have the same value of $m\mu/I$, and, for integral I, one energy level has no magnetic component in the z direction.

It should be noted that the size of the excess low-energy population is proportional to B_0. For this reason the magnetic effect of the nuclei and therefore their signal intensity increases as the strength of the magnetic field is increased. Temperature also has an important effect, and the value of $2\mu B_0/kT$ is increased to 7.2×10^{-5} at 250 K or decreased to 4.9×10^{-5} at 370 K at 9.39 T. This can produce detectable changes in signal strength relative to background noise in variable-temperature experiments.

We have so far built up a picture of the nuclei in a sample polarized with or against the magnetic field and lying at an angle θ to it. The total angular momentum (i.e. the length of the vectors of Fig. 1.1) is $[I(I + 1)]^{1/2}$ and the angle θ is then given by

$$\cos \theta = m/[I(I + 1)]^{1/2}$$

This angle can also be calculated classically by considering the motion of a magnet of moment μ in an applied magnetic field. It is found that the magnet axis becomes inclined to the field axis and wobbles or precesses around it. The magnet thus describes a conical surface around the field axis. The half-apex angle of the cone is equal to θ and the angular velocity around the cone is γB_0, so that the frequency of complete rotations is $\gamma B_0/2\pi$, the nuclear resonant frequency (Fig. 1.2a). This precession is known as the Larmor precession.

For an assembly of nuclei with $I = 1/2$ there are two such precession cones, one for nuclei with $m = +1/2$ and one for $m = -1/2$ and pointing in opposite directions. It is usual, however, to consider only the precession cone of the excess low-energy nuclei, and this is shown in Fig. 1.2b, which represents them as spread evenly over a conical surface and all rotating with the same angular velocity around the magnetic field axis, which is made the z axis. Since the excess low-energy nuclear spins all have components along the z axis pointing in the same direction, they add to give net magnetization M_z along the z axis. Individual nuclei also have a component μ_{xy} transverse to the field axis in the xy plane. However, because they are arranged evenly around the z axis, these components all average to zero, i.e. $M_x = M_y = 0$. The magnetism of the system is static and gives rise to no external effects other than a very small, usually undetectable, nuclear paramagnetism due to M_x.

In order to detect a nuclear resonance we have to perturb the system. This is done by applying a sinusoidally oscillating magnetic field along the y axis with frequency $\gamma B_0/2\pi$. This can be thought of as stimulating both absorption and emission of energy by the spin system (i.e. as stimulating upward and downward spin transitions), but resulting in net absorption of energy, since more spins are in the low-energy state and are available to be promoted to the high-energy state. The perturbing field is generated by passing a radiofrequency (RF) alternating current through a coil wrapped around the sample space.

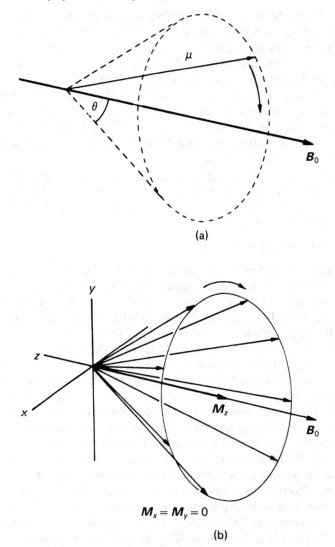

(a)

(b)

Fig. 1.2   Freely precessing nuclei in a magnetic field B_0. (a) Larmor precession of a single nucleus (b) The excess low-energy nuclei in a sample. The nuclear vectors can be regarded as being spread evenly over a conical surface. They arise from different atoms but are drawn with the same origin.

Classically we can analyse the oscillating magnetic field into a superposition of two magnetic vectors rotating in opposite directions. These add at different instants of time to give a zero, positive or negative resultant (Fig. 1.3). The vector B_1, which is rotating in the same sense as the nuclei (Fig. 1.4), is stationary relative to them, since we have arranged that it should have the same angular

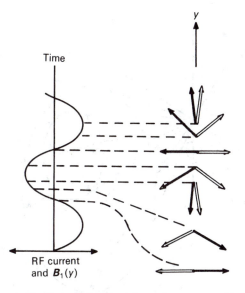

Fig. 1.3   The RF current in the coil in Fig. 1.4 produces an oscillating magnetic field along the y axis. This can be equally well regarded as being composed of two rotating magnetic vectors rotating in opposite directions, here shown as black and white arrows, whose resultant is the oscillating magnetic field. One of these vectors will rotate in the same sense as the nuclear precession and is conventionally called B_1.

velocity. The nuclei thus precess around B_1 also and are displaced from the main field axis. Since the nuclear moments do not now all have the same component in the xy plane, they no longer cancel in the xy direction and there is resultant magnetization M_{xy} transverse to the main field and rotating at the nuclear precession frequency. The magnetism of the system is no longer static and the rotating vector M_{xy} will induce a radiofrequency current in a coil placed around the sample, which we can detect, provided pick-up of B_1 can be avoided.

If B_1 is of large amplitude, then the nuclei swing very rapidly around it and the M_{xy} component of the magnetization, and so the current induced in the coil increases and reaches a maximum when the nuclear magnets have precessed 90° around B_1 (Fig. 1.5). When B_1 is cut off, of course, this precession stops. In practice, it is possible to cause a 90° precession in very short times of between 2 and 200 μs (1 μs = 10^{-6} s). The coil then detects the ensuing nuclear signal with no interference from B_1, though the signal intensity diminishes to zero as the system returns to equilibrium with $M_{xy} = 0$. An output will be observable in general for between 10 ms and 10 s.

In fact, we are not particularly concerned with the behaviour of individual nuclei, but rather with that of the resultant magnetization, and Fig. 1.5 is unnecessarily complex. It is usual to use the much simpler type of diagram shown in Fig. 1.6, which shows only the overall nuclear magnetization.

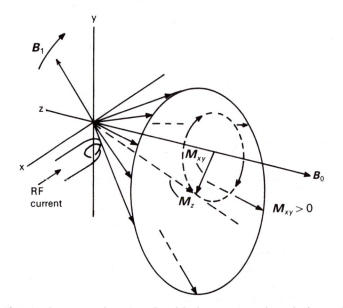

Fig. 1.4   If a rotating magnetic vector B_1 with the same angular velocity as the nuclei is now added to the system, the nuclei will also tend to precess around B_1, and this causes the cone of vectors to tip and wobble at the nuclear precession frequency. The resulting rotating vector M_{xy} in the xy plane can induce a current in the coil that is wound around the sample.

If B_1 does not precess at the same frequency as the nuclei, then the nuclear precession around B_1 is always changing direction and so M_{xy} can never become significant. It is this feature, whereby the signal is obtained from all the excess low-energy nuclei acting in concert and only at a single frequency, that gives to the technique its name of 'resonance spectroscopy'.

1.4   A basic NMR spectrometer

We are now in a position to understand the principles underlying the construction of an NMR spectrometer. The object is to measure the frequency of a nuclear resonance with sufficient accuracy. The instrument (Fig. 1.7) comprises a strong, highly stable magnet in whose gap the sample is placed and surrounded by transmitter–receiver coil C. The magnet may be of there types: a permanent magnet, an electromagnet or a superconducting (cryo)magnet for the highest field strengths. The field stability is ensured in the first case by placing the magnet in an isothermal enclosure whose temperature is controlled so as to

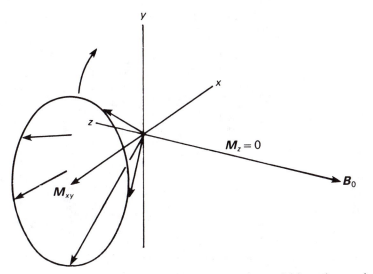

Fig. 1.5   A sufficiently long or powerful B_1 will turn the nuclei into the xy plane and the whole of the nuclear magnetization then contributes to the signal picked up by the coil in Fig. 1.4 and the output is at a maximum. B_1 is thus applied in the form of a pulse, and a pulse that has the effect illustrated is known as a 90° pulse. The spins follow a spiral path away from the B_0 axis during the pulse.

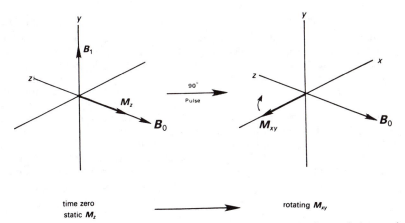

Fig. 1.6   A simplified form of Fig. 1.5 in which only the behaviour of the resultant nuclear magnetization is depicted. Note that we have frozen the position of B_1 by allowing the laboratory frame of reference to rotate with the nuclear precession. This is known as the rotating frame of reference, or simply as 'the rotating frame'. It permits us to see that happens to the magnetization in the presence of B_1 by eliminating the otherwise spiral paths that occur in normal space.

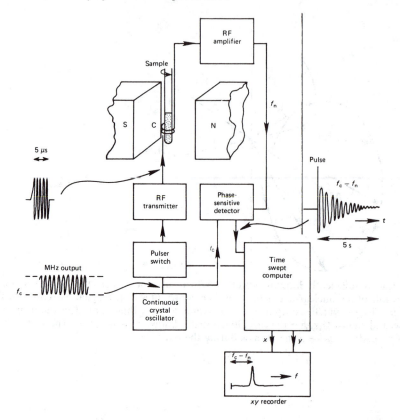

Fig. 1.7   A basic Fourier transform spectrometer. The $5\,\mu$s radiofrequency (RF) pulse tips the nuclei in the sample by $90°$ provided their frequency lies within the bandwidth of the pulse. (The pulse switching in effect converts the monochromatic crystal frequency f_c into a band of frequencies of width $1/t$, where t is the pulse length. In this case the bandwidth is $200\,000\,\mathrm{Hz}$.) The nuclear output signal will be at a frequency f_n close to f_c and the difference frequency $f_c - f_n$ is obtained at the output of the phase-sensitive detector. The computer collects the output and then calculates the frequency of resonance relative to f_c. The resonance frequency has limited definition (i.e. it has width), which is related to the time for which the output signal persists. Note that a time-dependent output is converted to a frequency-dependent output for analysis. Note also the very different timescales for the RF pulse and the output, here given the arbitrary lengths of $5\,\mu$s and $5\,$s. The computer and pulser have to be linked in some way to synchronize pulse timing and data collection.

drift not more than $10^{-4}\,$K per hour. An extremely stable power supply is used to energize the electromagnet, and then an auxiliary coil is used to detect remaining field variations and so to provide correcting adjustments. The superconducting magnet needs no current supply once the field is set up and so is inherently stable, though correcting circuits are still necessary. The magnetic field at the sample also inevitably varies throughout the bulk of the sample (i.e. the field is non-homogeneous), so that the signal frequency is not well defined. A further set of coils, known as shim coils, is placed around the sample in order to counteract these variations or field gradients and render the field as perfectly homogeneous as possible. The shim coils are not shown. Remaining inhomogeneities are minimized by spinning the sample tube about its long axis so that the sample molecule experience average fields. Very well defined frequencies, and so excellent resolution of close, narrow resonances, are obtained in this way. The B_1 field is produced by a gated (switched input) power amplifier driven by a stable, crystal-controlled continuous oscillator. Because the B_1 pulse is very short, its frequency is less well defined than that of the monochromatic crystal oscillator and has a bandwidth of $1/t$ Hz (t is the length of the pulse in seconds). It thus does not need to be exactly in resonance with the nuclei and indeed can cause simultaneous precession of nuclei with different frequencies. The nuclear signals are then amplified and detected in a device which compares them with the crystal oscillator output (B_1 carrier f_c) and gives a low-frequency, time-dependent output containing frequency, phase and amplitude information. This output is digitized and collected in a computer memory for frequency analysis using a Fourier transform program, and the spectrum that results (a spectrum is a function of frequency) can be output to an xy recorder and the resonance frequencies listed on a printer.

Not shown is a parallel spectrometer, which can detect deuterium in the sample and use this signal to stabilize the magnetic field. This is a field-frequency lock, which will be described in detail later.

This type of spectrometer is known as a Fourier transform (FT) spectrometer. It is also possible to obtain spectra from the more receptive nuclei using a much simpler system, which dispenses with the pulser and computer and produces the recording of the spectrum directly. This is the continuous-wave (CW) spectrometer. All early high-resolution spectrometers were of this type, and indeed the modern FT instruments were developed from these.

Questions

1.1. A spectrometer operating at a fixed magnetic field is set up to observe the nucleus ^{13}C at a frequency of 25 000 000 Hz (25 MHz). The sample examined contains two resonances at 25 000 250 Hz and 25 001 000 Hz respectively. What frequencies will be present at the output to the computer digitizer. What would these be if the spectrometer frequency were changed to 25 001 250 Hz?

1.2. A particular apparatus used to obtain ^{13}C spectra at 25 MHz has a 90° pulse length of 100 μs. Calculate the bandwidth (or frequency coverage) of this pulse. It is desired to obtain spectra over a range of 5000 Hz. Is the bandwidth sufficient for this? If we were able to increase the magnetic field of the spectrometer so that the ^{13}C operating frequency became 125 MHz (the frequency range needed would be 25 000 Hz), would the 100 μs pulse still be of sufficiently wide coverage? What is the maximum pulse angle that could be tolerated if the whole spectral range were to be stimulated uniformly?

1.3. In which of the following cases are M_z and/or M_{xy} zero: (a) following a 45° pulse, (b) following a 90° pulse, and (c) following a 180° pulse?

2 The magnetic field at the nucleus: nuclear screening and the chemical shift

2.1   Effects due to the molecule

So far we have shown that a single isotope gives rise to a single nuclear magnetic resonance in an applied magnetic field. This really would be of little interest to the chemist except for the fact that the magnetic field at the nucleus is never equal to the applied field, but depends in many ways upon the structure of the molecule in which the atom carrying the nucleus resides.

The most obvious source of perturbation of the field is that which occurs directly through space due to nuclear magnets in other atoms in the molecule. If such nuclei have high magnetic moments, which means generally ^{1}H or ^{19}F, then in the solid state this interaction results in considerable broadening of the resonance, which obscures much information. However, in the liquid state, where the molecules rotate rapidly and randomly, the direct nuclear fields fluctuate wildly in both intensity and direction and have an average value that is exactly zero. The resonances are thus narrow and may show much structure. Thus spectroscopy of the liquid state has been a major preoccupation of chemists.

Since the magnetic nuclei do not directly perturb the field at the nucleus, we have therefore to consider the effect that the electrons in the molecule may have. We will concern ourselves only with diamagnetic molecules at this stage and will defer till later discussion of paramagnetic molecules possessing an unpaired electron. When an atom or molecule is placed in a magnetic field the field induces motion of the electron cloud such that a secondary magnetic field is set up. We can think of the electrons as forming a current loop as in Fig. 2.1 centred on a positively charged atomic nucleus. The secondary field produced by this current loop opposes the main field at the nucleus and so reduces the nuclear frequency. The magnitude of the electronic current is proportional to B_0 and we say that the nucleus is screened (or shielded) from the applied field by its electrons. This concept is introduced into equation (1.2) relating field and nuclear frequency by the inclusion of a screening constant σ:

$$v = \frac{\gamma B_0}{2\pi}(1 - \sigma) \tag{2.1}$$

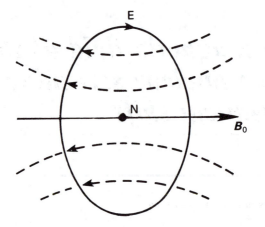

Fig. 2.1   The motion of the electronic cloud E around the nucleus N gives rise to a magnetic field, shown by dashed lines, which opposes B_0 at the nucleus.

σ is a small dimensionless quantity and is usually recorded in parts per million (ppm). The screening effect is related to the mechanism that gives rise to the diamagnetism of materials and is called diamagnetic screening.

The magnitude of the effect also depends upon the density of electrons in the current loop. This is a maximum for a free atom where the electrons can circulate freely, but in a molecule the free circulation around an individual nucleus is hindered by the bonding and by the presence of other positive centres, so that the screening is reduced and the nuclear frequency increased. Since this mechanism reduces the diamagnetic screening it is known as a paramagnetic effect. This is unfortunately a misleading term and it must be emphasized that it does not imply the presence of unpaired electrons. As used here, it merely indicates that there are two contributions to σ, the diamagnetic term σ_d and the paramagnetic term σ_p and that these are opposite in sign. Thus

$$\sigma = \sigma_d + \sigma_p \qquad (2.2)$$

Since the magnitude of σ_d depends upon the density of circulating electrons, it is common to find in the literature discussion of the effect of inductive electron drifts on the screening of nuclei. The screening of protons in organic molecules, for instance, depends markedly on the substituents, and good linear correlations have been found between screening constants and substituent electronegativity, thus supporting the presence of an inductive effect. Currently, however, it is believed that most of these variations originate in the long-range effects to be described below and that the contribution of inductive effects is small, at least for σ-bonded systems.

The magnitude of the paramagnetic contribution σ_d is zero for ions with spherically symmetric s states but is substantial for atoms, particularly the heavier ones, with many electrons in the outer orbitals involved in chemical bonding. It is determined by several factors.

1. The inverse of the energy separation ΔE between ground and excited electronic states of the molecule. This means that correlations are found between screening constants and the frequency of absorption lines in the visible and ultraviolet.
2. The relative electron densities in the various p orbitals involved in bonding, i.e., upon the degree of asymmetry in electron distribution near the nucleus.
3. The value of $\langle 1/r^3 \rangle$, the average inverse cube distance from the nucleus to the orbitals concerned.

In the case of hydrogen, for which there are few electrons to contribute to the screening, and for which ΔE is large, σ_d and σ_p are both small and we observe only a small change in σ among its compounds, most of which fall within a range of 20×10^{-6} or 20 ppm. In the case of elements of higher atomic number, ΔE tends to be smaller and more electrons are present, so that, while both σ_d and σ_p increase, σ_p increases disproportionately and dominates the screening. Thus changes in σ_p probably account for a major part of the screening changes observed for boron in its compounds (a range of 200 ppm); and σ_p almost certainly predominates for fluorine (where the range is 1300 ppm) or for thallium (where it is 5500 ppm). We see that the changes observed for the proton are thus unusually small.

In the case of screening of the fluorine nucleus, the values for fluorine compounds are known relative to the bare, unscreened fluorine nucleus. In certain of its compounds, fluorine is less screened even than the bare nucleus, some examples beingg F_2, UF_6 or FOOF. Presumably this arises because of electronic circulation near to, but not centred upon, the nucleus.

The observable changes in screening of each nucleus do not increase continuously with atomic number but exhibit a periodicity, increasing steadily along each period but then falling markedly at the start of the next. The behaviour down each group is similar, as is shown in Fig. 2.2. This periodicity follows closely the values of $\langle 1/r^3 \rangle$ for each element.

Because we are dealing with the effects of electronic circulation in a three-dimensional molecule relative to a unidirectional magnetic field, it is easy to see that the nature of the circulation will change as the orientation of the molecule changes in the magnetic field. Thus the screening at any instant is a function of the attitude of the molecule relative to $\boldsymbol{B}_0$. These orientational effects are fully described by the screening tensor, which has nine components, though only three influence the observed screening. These, the diagonal components of the tensor, are called σ_{11}, σ_{22} and σ_{33} and all three are required if the nucleus is in a site with no symmetry. If the site is axially symmetric, then only two values are needed to describe the screening since $\sigma_{11} = \sigma_{22}$. These components are then called $\sigma_\perp$ and σ_{33} is denoted $\sigma_{\parallel}$. This is called the screening anisotropy, and the differences between the values of the components can be substantial.

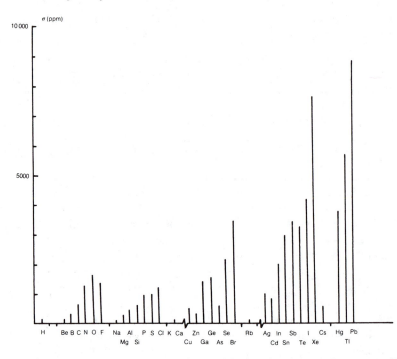

Fig. 2.2 Ranges of screening constants for nuclei of main-group and post-transition elements. It should be remembered that, while the ranges shown reflect the periodicity of the $\langle 1/r^3 \rangle$ term, they also are influenced by how many compounds of a given element have been measured and, indeed, by the extent of its chemistry. (After Jameson and Mason (1987) *Multinuclear NMR*, Plenum, with permission.)

Screening anisotropy of a nucleus can be observed by taking its spectrum from a powdered solid where the solid particles all have different orientations. Two types of spectrum are shown diagrammatically in Fig. 2.3, and the values of the tensor components are given by the discontinuities on the curves. The effect is another source of line broadening in solids but also allows a full description of the screening mechanism to be obtained. In liquid samples the isotropic rotation of the molecules produces an average σ where $\sigma = (\sigma_{11} + \sigma_{22} + \sigma_{33})/3$ and the lines become narrow. The high-resolution spectrum that is thus obtained is generally more useful to the chemist, but it should not be forgotten that some fundamental information is lost in the process.

The variations in screening among the compounds of a given element depend upon all the factors summarized above, but in a way that is very difficult to separate into dominance by any particular influence. For instance, it seems certain that increased charge density upon an atom results in increased screening of its nucleus but that this occurs principally because the $\langle 1/r^3 \rangle$ term is

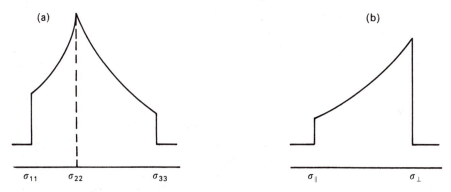

(a)

σ_{11} σ_{22} σ_{33}

(b)

$\sigma_{\parallel}$ $\sigma_{\perp}$

Fig. 2.3   Representations of the powder spectra of two different solid samples: (a) of a nucleus in an asymmetric environment and (b) of a nucleus in an axially symmetric environment. Broadening due to through-space magnetic interaction with other nuclei in the samples is avoided by choosing samples where protons, for example, are relatively distant.

increased; in other words the change in orbital radius determines the change in screening. Two examples serve to emphasize the complexity of the situation. As expected from simple considerations of charge density, the ^{14}N nucleus in NH_3 is 25 ppm more shielded than in NH_4^+, the influence of the apparent change in coordination number being slight since NH_3 has an electron lone pair as effective fourth ligand. In contrast, the ^{14}N nucleus in

is 100 ppm more shielded than that in pyridine

In this case the nitrogen is part of a π system and protonation removes an accessible low-energy transition and so modifies the ΔE term. Oxidation state also effects screening via electron density changes. The influence of electron imbalance in the bonds is manifested by the common effect in which increased screening follows increases in coordination number; for example ^{31}P screening in $PCl_3 < PCl_4^+ < PCl_5 < PCl_6^-$, and several similar trends will be observed in the scales of Fig. 2.12. It must, however, be emphasized that this effect is not universally true and other considerations operate in some cases. In all the normal cases the increase in coordination number can be regarded as leading to an increase in local symmetry and so a more balanced structure. Electron imbalance is also related to bond ionicity, π-bond order or s character and so to bond angles. Thus the ^{31}P nucleus is 125 ppm more screened in trimethylphosphine,

PMe$_3$, than in the sterically crowded tri-t-butylphosphine, PBu$_3^+$, because of the increased C–P–C bond angles in the latter.

Substituents also exhibit a marked effect upon nuclear screening, which often correlates with changes in substituent electronegativity. However, the direction of the correlation is not uniform, and screening may increase or decrease with increase in electronegativity depending upon the sequence of ligands chosen. Again, several effects can operate simultaneously. For instance, the ^{27}Al nucleus is increasingly screened in the series of tetrahalo anions AlCl$_4^-$ < AlBr$_4^-$ < AlI$_4^-$. This is known as the normal halogen dependence, since it is in the opposite sense for certain transition metals. The change in screening between chloride and bromide (22 ppm) is less than that between bromide and iodide (47 ppm). The changes are ascribed first to the nephelauxetic effect of the halogen, the larger halogens expanding the electron orbitals on the Al and so decreasing the $\langle 1/r^3 \rangle$ term and so σ_p. In addition, the heavier halogens produce the heavy-atom screening effect, which is proportional to Z^4, where Z is the nuclear charge of the halogen, and this enhances the apparent nephelauxetic effect and explains the bigger increase in screening between the bromide and iodide. The heavy-atom effect operates via a complex relativistic spin–orbit coupling mechanism.

Usually the contributions to σ_d and σ_p for a nucleus are considered only for the electrons immediately neighbouring, or local to, that nucleus. More distant electrons give rise to long-range effects on both σ_d and σ_p, which are large but cancel to make only a small net contribution to σ. It is therefore more convenient to separate the long-range effects into net contributions from different, quite localized, parts of the rest of the molecule. Two types of contribution to screening can be recognized, and though they are small they are of particular importance for the proton resonance.

2.1.1   Neighbour anisotropy effects

We have already mentioned that in liquid samples, owing to the rapid and random motion of the molecules, the magnetic fields at each nucleus due to all other magnetic dipoles average to zero. This is only true if the magnet (e.g. a nucleus) has the same dipole strength whatever the orientation of the molecule relative to the field direction. If the source of magnetism is anisotropic and the dipole strength varies with orientation in the applied field, then a finite magnetic field appears at the nucleus.

Such anisotropic magnets are formed in the chemical bonds in the molecule, since the bonding electrons support different current circulation at different orientations of the bond axis to the field. The result is that nuclei in some parts of the space near a bond are descreened while in other parts the screening is increased. Figure 2.4 and 2.5 show the way screening varies around some bonds.

A special case of anisotropic screening where the source of the anisotropy

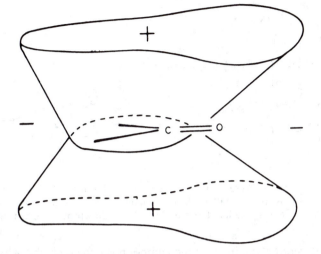

Fig. 2.4 Screened and descreened volumes of space around a carbonyl bond. The sign + indicates that a nucleus in the space indicated would be more highly screened. The magnitude of the screening falls off with increasing distance from the group and is zero in the surface of the solid figure.

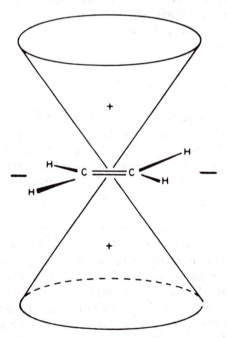

Fig. 2.5 Screened and descreened volumes of space around a carbon–carbon double bond. The significance of the signs is the same as in Fig. 2.4. The cone axis is perpendicular to the plane containing the carbon and hydrogen atoms.

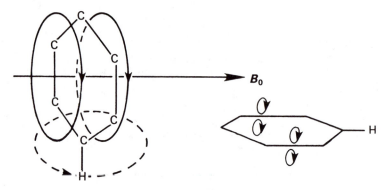

Fig. 2.6   Ring current descreening in benzene. The area of the ring current loops due to the π electrons is much smaller when the plane of the ring lies along the field axis and the reaction field is smaller. The reaction field is shown by a dashed line.

is clearly evident occurs in aromatic compounds, which exhibit what is called ring current anisotropy. The benzene structure, for instance, can support a large electronic ring current around the conjugated π-bond system when the plane of the ring is transverse to the field axis but very little when the ring lies parallel to the field axis. This results in large average descreening of benzene protons since the average secondary magnetic field, which must oppose the applied field within the current loop, acts to increase the field outside the loop in the region of the benzene protons (Fig. 2.6).

2.1.2   Through-space electric field effects

Molecules that contain electric dipoles or point charges possess an electric field whose direction is fixed relative to the rest of the molecule. Such electric fields can perturb the molecular orbitals by causing electron drifts at the nuclei in the bond directions and by altering the electronic symmetry. It has been shown that the screening σ_E due to such electric fields is given by

$$\sigma_E = -AE_z - BE^2 \tag{2.3}$$

where A and B are constants, $A \gg B$, E_z is the electric field along a bond to the atom whose nuclear screening we require and E is the maximum electric field at the atom. The first term produces an increase in screening if the field causes an electron drift from the bond onto the atom and a decrease if the drift is away from the atom. The second term leads always to descreening. It is only important for proton screening in the solvation complexes of highly charged ions where E can be very large, though it is of greater importance for the nuclei of the heavier elements.

The electric field effect is, of course, attenuated with increasing distance. It is an intramolecular effect since the effect of external fields, for which the BE^2

term can be neglected, averages to zero as the molecule tumbles and E continually reverses direction along the bond.

The descreening of protons that occurs in many organic compounds containing electronegative substituents X probably occurs because of the electric field set up by the polar C–X bond. This will also increase with X electronegativity and produce a similar result to an inductive electron drift. The effect typically produces proton descreening in molecules containing keto, ester, or ether groups and in halides, some values being given below:

Molecule	CH_3Cl	$CH_3–C(O)O–CH_3$	$(CH_3)_2CO$	$(CH_3)_2O$
ppm less screened than methane	2.83	1.78 3.44	1.86	3.01

The closer the protons are to the bond generating the electric field, then the more they are descreened. This also shows up as a fall-off in descreening along an alkyl chain for the protons further away from the substituent, and, for instance, in 1-chloropropane (n-propyl chloride) the comparable descreening figures are α-CH_2 3.24 ppm, β-CH_2 1.58 ppm and CH_3 0.83 ppm.

2.2   Isotope effects

Because the bonds in molecules are not rigid fixed entities but have dimensions determined by vibrational phenomena, the substitution of an atom by one of its isotopes of different mass alters the vibrational energies in the bonds to that atom and so alters the electron distribution about it. This necessarily implies changes in nuclear screening following the substitution, both near the substitution site and at some distance from it. Such changes are small but measurable in many cases, and provide useful spectroscopic data. It is necessary to distinguish between primary isotope effects and secondary isotope effects. The former are the effects experienced by the isotopically substituted nucleus itself. For instance, for $^{14}N/^{15}N$ substitutions, the change in ^{14}N screening between $^{14}NH_3$ and CH_3 $^{14}NO_2$ is not the same as that for ^{15}N between $^{15}NH_3$ and CH_3 $^{15}NO_2$, given identical conditions. Clearly, such changes are not easy to measure and for many elements are within experimental error, so that primary effects will not be considered further. The secondary effects are those observed on the nuclei in the rest of the molecule; in the above example the proton screening change, say, between $^{14}NH_3$ and $^{15}NH_3$. The trends observed are as follows:

1. The isotope shift is greatest when the substitution causes the greatest fractional change in mass; thus $^2D/^1H$ substitution is the most effective.
2. Screening is greatest for nuclei near the heavier isotope. This is not always true and in some cases the opposite holds. The nearer the nucleus observed is to the substituted site, the greater is the effect.

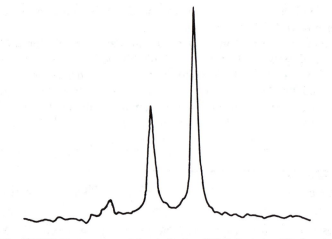

Fig. 2.7   The ^{15}N NMR spectrum of the nitrite ion, NO_2^-, in water. The ion was enriched to 95% in ^{15}N and 77% in ^{18}O. The three resonances arise from ions of isotopic composition $^{15}N^{16}O_2$, $^{15}N^{16}O^{18}O$ and $^{15}N^{18}O_2$ in the concentration ratios 6:33:61. Replacement of ^{16}O by ^{18}O causes an upfield shift of 0.138 ppm. (From Van Etten and Risley (1981) *J. Am. Chem. Soc.*, **103**, 5634; copyright (1981) American Chemical Society, reprinted with permission.)

3. The magnitude of the effect reflects the overall range of screening experienced by the observed nucleus (cf. Fig. 2.2).
4. The effect of multiple isotopic substitutions is additive, or approximately so.

Thus the ^{15}N signal obtained from the nitrite ion, $^{15}NO_2^-$, in water and which has had a partial substitution of ^{16}O by ^{18}O, fully randomized over the two positions, consists of three signals as shown in Fig. 2.7. This spectrum shows the additive effect and the greater screening due to the ^{18}O. It also proves unequivocally that the nitrite anion contains two oxygen atoms. Similarly, the ^{31}P spectrum of partially ^{18}O substituted PO_4^{3-} is an equally spaced quintet with spacing of 0.0206 ppm and proves that the orthophosphate contains four oxygen atoms.

This property of partial isotopic substitution provides an almost digital technique for measuring numbers of exchangeable atoms in a structure. It is, for instance, now possible to count the number of water molecules in aqua complexes such as $Al(H_2O)_6^{3+}$. Analytical methods never give whole numbers, and while the hexaaqua cation is known to exist in solids from X-ray structural analysis, it is possible to argue that the hydration number may not be so definite in solution. If $Al(H_2O)_6(ClO_4)_3$ is dissolved in acetone, it is possible to observe the water proton signal and note that this is highly descreened relative to free water. This is a consequence of the strong electric field of the cation. If a proportion of the water is replaced by D_2O, then the complex will contain H_2O, HOD and D_2O. The signal of the HOD protons shows a strong isotope effect, through this is abnormal as they are less screened even than the H_2O.

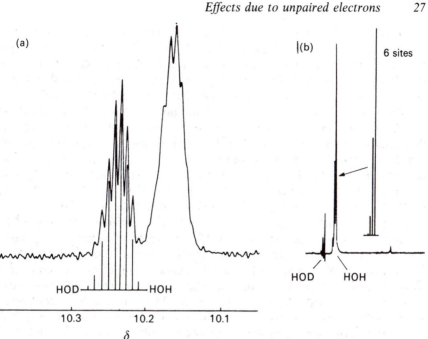

Fig. 2.8 (a) The 400 MHz proton spectrum of the water complexed to the cation $Al(H_2O)_6^{3+}$ in ($^2H_6$)acetone (deuterioacetone or acetone-d_6) taken at $-30°C$. The complex had been partially deuteriated so as to contain 35% ^2H. The resonance at 10.23 ppm is due to all the HOD molecules in the complex and that at 10.17 ppm is due to all the HOH molecules. The fine structure arises because different molecules have different total numbers of ^2H, each giving a smaller isotope effect due to the more distant substitution. The stick diagram gives the calculated intensities obtained from the deuterium content and assuming completely random distribution throughout the sample. (b) The equivalent spectrum obtained from $Be(H_2O)_4^{2+}$, demonstrating the marked difference between different numbers of exchangeable sites. (After Akitt *et al.* (1986) *J. Chem. Soc., Chem. Commun.*, 1047; (1988) *Bull. Soc. Chim. Fr.*, 466, with permission.)

This probably arises because the deuterium substitution shortens the Al–O bond in that molecule and so increases the electric field effect. More importantly, the two proton resonances (HOD and HOH) show fine structure as in Fig. 2.8. This arises because of long-distance isotope effects between a given HOD and the other ten replaceable sites in the complex. A 13-line pattern is theoretically possible, though some lines are too weak to detect, and one has to calculate an intensity distribution from the known level of deuteration.

2.3   Effects due to unpaired electrons

The electron (spin $= 1/2$) has a very large magnetic moment and if, for instance, paramagnetic transition-metal ions are present in the molecule, large effects are observed. The NMR signal of the nuclei present may be undetectable, but under certain circumstances, when the lifetime of the individual electron in each spin state is short, so that its through-space effect averages to near zero, NMR spectra can be observed. The screening constants measured in such systems, however, cover a very much larger range than is normal for the nucleus, and this arises because the electronic spins can be apparently delocalized throughout a molecule and appear at, or contact, nuclei. The large resonance displacements that results are known as contact shifts and the ligands in certain transition-metal–ion complexes exhibit proton contact shifts indicating several hundred ppm changes in σ. In addition, if the magnetic moment of the ion is anisotropic, one gets a through-space contribution to the contact interaction similar to the neighbour anistropy effect, and this is called a pseudo-contact shift.

Such effects are particularly marked with the lanthanide cations, whose complexes are commonly used to simplify the spectra of organic compounds. These substances are called shift reagents and consist of a lanthanide element (Eu, Dy, Pr, or Yb) complexed with an organic ligand chosen, among other things, to make the complex soluble in organic solvents so that it can be co-dissolved with the compound to be investigated. They are octahedral complexes, but the lanthanides are capable of assuming higher coordination numbers than six, so that if the organic molecule possesses a suitable donor site such as O

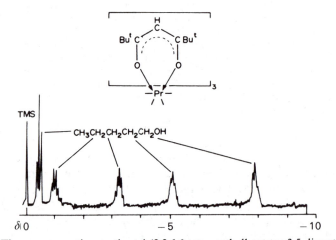

Fig. 2.9   The paramagnetic complex tris(2,2,6,6-tetramethylheptane-3,5-dionato)praseodymium has three bulky tridentate ligands surrounding the metal in such a way that it is octahedrally coordinated by oxygen. Only one of the ligands is shown. Molecules with lone pairs of electrons also approach the paramagnetic metal since the lanthanide can attain coordination numbers greater than six. Their protons can thus interact with the unpaired electrons on the metal and undergo pseudo-contact shifts. In the case of n-pentanol, a first-order spectrum is obtained. (Bu^t represents (CH$_3$)$_3$ C-, i.e. the t-butyl group. This type of ligand is used because it renders the ions soluble in organic solvents). (After Briggs *et al.* (1970) *J. Chem. Soc., Chem. Commun.*, 749, with permission.)

or N, it can interact with the shift reagent. This produces a pseudo-contact shift of all the proton and carbon nuclei in the molecule, which can represent very large changes in screening. For instance, the normal proton spectrum of pentanol consists of three signals that are all descreened relative to the signal of tetra-methylsilane (TMS), which as we shall see is usually used as a marker. If a praseodymium (Pr) complex is added, all the protons become more highly screened and the signals are now on the other side of the marker and, more importantly, are all well separated (Fig. 2.9). The fine structure is due to spin–spin coupling, which we will meet in the next chapter. If the organic molecule is rigid, then the magnitude of the shifts can be used to calculate its geometry using an interative method that optimizes both geometry and position of the lanthanide.

A second use of these reagents is in determining chirality or optical purity of organic substrates. This is achieved by preparing chiral shift reagents, i.e. ones in which the ligands are themeselves chiral. Such reagents produce slightly different screening effects in substrates of different handedness. It is, for instance, possible to obtain separate signals for the two optical isomers of $C_5H_{11}CHDOH$, which can be prepared using optically active reducing agents. Figure 2.10 shows the very small screening separation between the signals of the two forms of some 0.07 ppm.

2.4   The chemical shift

So far we have talked only in terms of the changes in the screening of nuclei in different environments. It is, however, more usual to describe these changes as 'chemical shifts' although the word 'screening' will still be encountered from time to time. We have already seen in Fig. 2.9 that, in order to emphasize the screening changes that can take place in the presence of a shift reagent, it is useful to use a marker signal of an inert substance (TMS). This technique is general and changes in screening relative to a suitable marker give calibrated chemical shifts.

Thus if we have two nuclei in different environments with screening constants σ_1 and σ_2, then the two nuclear frequencies in a given magnetic field B_0 are

$$v_1 = \frac{\gamma B_0}{2\pi}(1 - \sigma_1) \tag{2.4a}$$

$$v_2 = \frac{\gamma B_0}{2\pi}(1 - \sigma_2) \tag{2.4b}$$

whence

$$v_1 - v_2 = \frac{\gamma B_0}{2\pi}(\sigma_2 - \sigma_1) \tag{2.5}$$

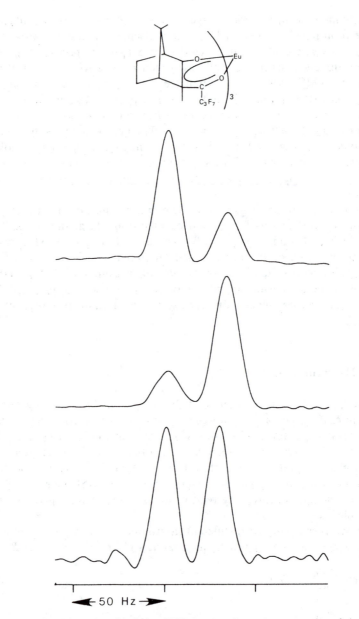

Fig. 2.10 The proton signals of the CHD protons in various samples of the optically active hexanol, $C_5H_{11}CHDOH$, made using an optically active reducing agent and obtained in the presence of a chiral shift reagent, in which the ligands are hexafluoropropyl camphorate (see formula). The lower trace shows the racemic mixture and the two upper traces are from materials made from reducing agents of opposite chirality. (Example supplied by B. E. Mann.)

We cannot measure the absolute value of B_0 accurately enough for this relation to be of use, so we eliminate field from the equation by dividing through by v_1 (equation (2.4a)). This gives us the frequency change as a fraction of v_1:

$$\frac{v_1 - v_2}{v_1} = \frac{\sigma_2 - \sigma_1}{1 - \sigma_1}$$

which since $\sigma_1 \ll 1$ reduces to

$$\frac{v_1 - v_2}{v_1} = \sigma_2 - \sigma_1 \tag{2.6}$$

Thus the fractional frequency change is the same as the difference in screening in the two nuclear environments. This is called the chemical shift and is given the symbol δ. Its value is expressed in parts per million (ppm). It can be determined with high accuracy since it is possible to resolve shifts of 0.001 ppm for spin-1/2 nuclei, or even less in favourable cases.

In order to establish a chemical shift scale for a given nucleus, it is necessary to choose some substance as a standard and define its chemical shift as zero. The usual, almost universal, standard for the three nuclei, the proton (^1H), carbon (^{13}C) and silicon (^{29}Si), is tetramethylsilane, $(CH_3)_4Si$, usually called TMS. It gives a narrow singlet resonance for protons, and for the other two nuclei if decoupling is used (see later for an explanation of this), which in all cases is outside the normal range of chemical shifts of the compounds studied. It is miscible with most organic solvents, it is inert and, being highly volatile, can easily be removed after measurements have been made.

In the case of ^1H and ^{13}C spectroscopy the resonances of interest come predominantly from nuclei that are less screened than is TMS, the main exceptions occurring where metal atoms are present in the compounds studied.

It is usual in recording spectra to depict the descreened region to the left-hand side of the spectrum with TMS to the right. The frequency then increases to the left of TMS since the magnetic fields at descreened nuclei are apparently higher and the nuclear precession frequency is higher. For historical reasons, however, this region is very commonly referred to as 'low field', the aptness of this name being apparent from equation (2.1). Both the names 'low field' and 'high frequency' are met in practice, the latter being the most logical since in a fixed-field instrument it is indeed the frequencies that change. The older name, though, does explain the apparently eccentric way in which the shifts are displayed. This is summarized in Fig. 2.11. Because we have chosen our standard arbitrarily, we also find that we have introduced sign into the scale. Thus in ^1H spectroscopy all those protons high field of TMS (low frequency) have negative shift values. This scale is called the δ scale, and it is important to note that the symbol δ may either refer to this scale or simply be used as short-hand for 'chemical shift'. In older work the sign convention may be reversed and an alternative TMS scale is also used, the τ scale, where $\tau = 10 - \delta$.

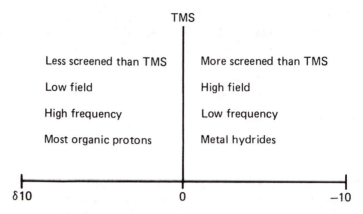

Fig. 2.11   Summary of proton chemical shift scales.

It is, of course, possible to use subsidiary standards, which are referred in turn to TMS. ^{1}H spectroscopy in aqueous solution thus often has recourse to either $(CH_3)_3Si(CH_2)_3COOH$ or $(CH_3)_4N^+$.

Calculation of a chemical shift is a sample matter. Modern spectrometers give the frequency separation of signals and it is only necessary to divide by the spectrometer operating frequency. If in a given solution the signal of benzene is 438 Hz higher in frequency than TMS and the spectrometer operating frequency is 60 MHz (i.e. $B_0 = 1.41$ T), then the chemical shift is

$$\delta = \frac{438}{60\ 000\ 000} \times 10^6 = 7.3 \text{ ppm}$$

Or, more easily remembered, it is the frequency difference in hertz divided by the operating frequency in megahertz. We see also that the frequency separations will be different in spectrometers operating at different frequencies. Thus in a 200 MHz spectrometer, the frequency separation above will be 1460 Hz.

The conventions adopted for other nuclei are less firm. The shifts are usually large, so that it is not quite so important to be able to compare different workers' results with high accuracy, and the standard substance is often chosen according to the dictates of convenience. The standard is assigned 0 ppm and the δ scales are then as in Fig. 2.12. An aqueous salt solution is often used as standard for groups 1, 2, 3, and 9: $^7Li^+$ (aq) for lithium, $^{27}Al(H_2O)_6^{3+}$ for aluminium and $^{35}Cl^-$ (aq) for chlorine, for instance. Some much used references are $(CH_3O)_3B$ and $(CH_3CH_2)_2O \rightarrow BF_3$ for ^{11}B spectroscopy; nitromethane, CH_3NO_2, or nitrate ion, NO_3^-, for ^{14}N and ^{15}N spectroscopy; H_2O for ^{17}O spectroscopy; 85% orthophosphoric acid, H_3PO_4, or occasionally P_4O_6 for ^{31}P spectroscopy; and the refrigerant $CFCl_3$ is commonly used as standard for ^{19}F work. This ^{19}F chemical shift scale is often called the ϕ scale. Other standards used for ^{19}F spectroscopy are hexafluorobenzene, C_6F_6, or trifluoroacetic acid.

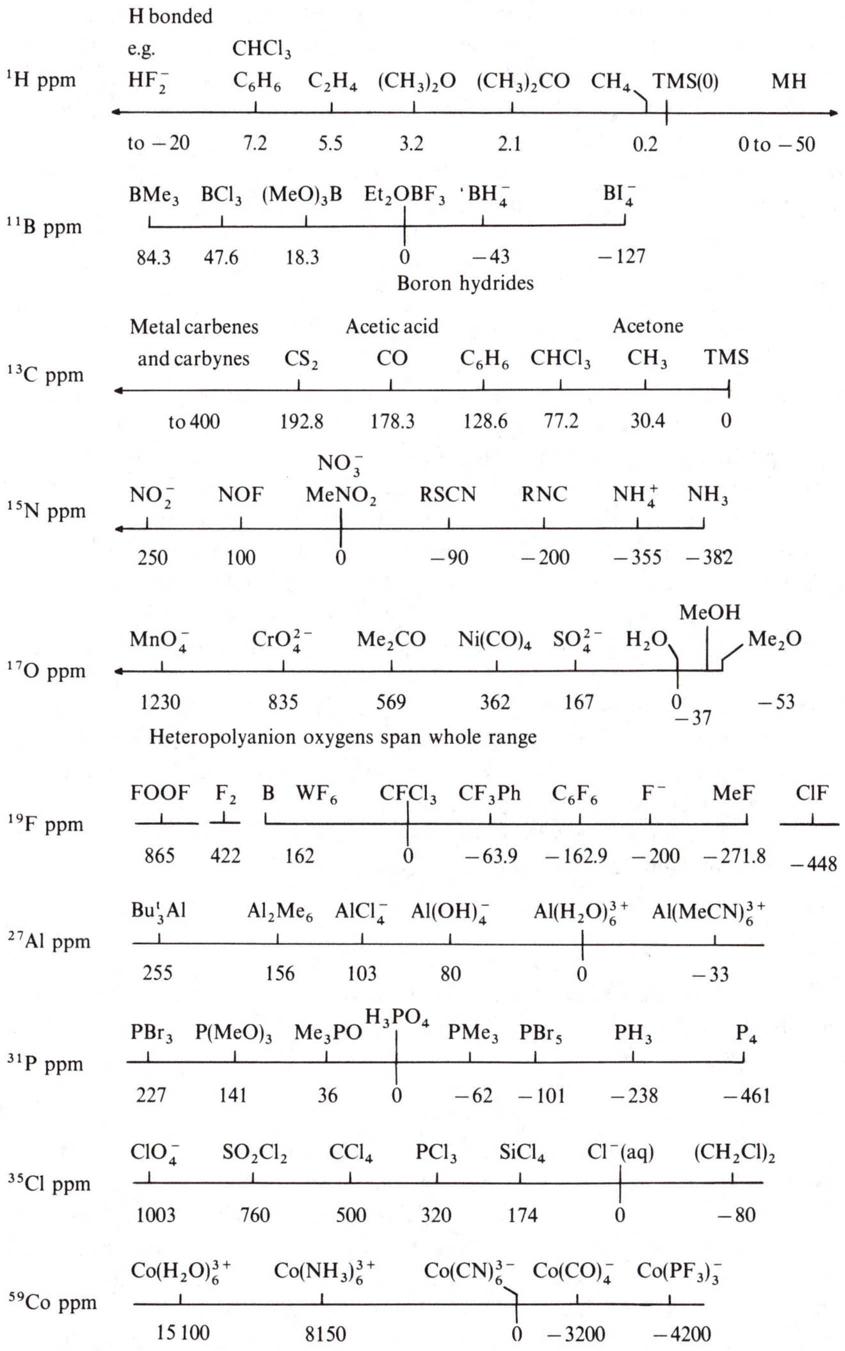

¹H ppm

H bonded
e.g.   CHCl₃

HF_2^- C_6H_6 C_2H_4 $(CH_3)_2O$ $(CH_3)_2CO$ CH_4 TMS(0)   MH

to -20 7.2   5.5   3.2   2.1   0.2   0 to -50

¹¹B ppm

BMe_3 BCl_3 $(MeO)_3B$ Et_2OBF_3 $'BH_4^-$ BI_4^-

84.3   47.6   18.3   0 -43 -127

Boron hydrides

¹³C ppm

Metal carbenes and carbynes CS_2 CO C_6H_6 CHCl₃ CH_3 TMS

Acetic acid   Acetone

to 400   192.8   178.3   128.6   77.2   30.4   0

¹⁵N ppm

NO_2^- NOF NO_3^- $MeNO_2$ RSCN   RNC NH_4^+ NH_3

250   100   0 -90 -200 -355 -382

¹⁷O ppm

MnO_4^- CrO_4^{2-} Me_2CO $Ni(CO)_4$ SO_4^{2-} H_2O MeOH Me_2O

1230   835   569   362   167   0 -37 -53

Heteropolyanion oxygens span whole range

¹⁹F ppm

FOOF F_2 B WF_6 $CFCl_3$ CF_3Ph C_6F_6 F^- MeF   ClF

865   422   162   0 -63.9 -162.9 -200 -271.8 -448

²⁷Al ppm

Bu_3^tAl Al_2Me_6 $AlCl_4^-$ $Al(OH)_4^-$ $Al(H_2O)_6^{3+}$ $Al(MeCN)_6^{3+}$

255   156   103   80   0 -33

³¹P ppm

PBr_3 $P(MeO)_3$ Me_3PO H_3PO_4 PMe_3 PBr_5 PH_3 P_4

227   141   36   0 -62 -101 -238 -461

³⁵Cl ppm

ClO_4^- SO_2Cl_2 CCl_4 PCl_3 $SiCl_4$ $Cl^-(aq)$ $(CH_2Cl)_2$

1003   760   500   320   174   0 -80

⁵⁹Co ppm

$Co(H_2O)_6^{3+}$ $Co(NH_3)_6^{3+}$ $Co(CN)_6^{3-}$ $Co(CO)_4^-$ $Co(PF_3)_3^-$

15 100   8150   0 -3200 -4200

Fig. 2.12   Various chemical shift scales and the chemical shifts of some compounds.

These and some other chemical shift scales are illustrated in Fig. 2.12, which shows the standards used (0 ppm) and the chemical shifts of some compounds, chosen to cover the full range of shifts for a given element rather than to pick out any particular trends with composition. For many of the heavier elements, the shift is very medium-sensitive and the spot value given is purely illustrative. For the ^{19}F scale, the shift marked B is that of the bare nucleus (189 ppm). Note also the very large chemical shift range for the transition-metal nucleus ^{59}Co.

2.5   Notes on sample preparation, standardization and solvent effects

An NMR experiment involves a highly sophisticated instrument capable of resolving resonances and making measurements to a few parts in 10^9. (This is equivalent to comparing the lengths of two steel rods each 1 km long to within one-thousandth of 1 mm). One must, therefore, accept the responsibility of preparing a sample that will not degrade the spectrometer performance. However, any sample placed in the magnet gap will distort the magnetic field. Fortunately, the distortion occurs externally to a cylindrical sample and the field remains homogeneous within it except at the ends, though its magnitude is changed by an amount that depends both on the shape of the sample and on the bulk magnetic susceptibility of the tube glass and of the sample itself. Imperfections in the glass, variations in wall thickness, variations in diameter, or curvature of the cylinder along its length all lead to degradation of the field homogeneity within the sample, with consequent line broadening. For this reason high-precision bore sample tubes are always used for NMR. Since solid particles distort the field around them, suspended solids must also be filtered from the liquid sample prior to measurement.

2.5.1   Standardization

We have shown that chemical shifts are invariably measured relative to a standard of some sort. There are three ways of standardizing a resonance, which are now given.

(a) Internal standardization

The standard substance is dissolved in the sample solution and its resonance appears in the spectrum. This method has the advantage that the magnetic field is exactly the same at sample and standard molecules. The standard must be chosen so as not to obscure sample resonances and also must be inert to the sample. Internal standardization is the method normally used and the techniques (b) and (c) below are used only in special cases.

The main disadvantage of the method is that weak interactions with the solvent produce small chemical shifts, which are difficult to predict and which reduce the accuracy of the measurements by an unknown amount. These shifts are said to arise from solvent effects.

(b) External standardization

The standard is sealed into a capillary tube that is placed coaxially within the sample tube. The main disadvantage of the method is that, since the volume magnetic susceptibilities of the sample and standard will differ by several tenths of a ppm, the magnetic fields in each will be different and a correction will have to be made for this. Since the volume susceptibilities of solutions are often not known, these must be measured, so that a single shift determination becomes quite difficult if accurate work is required. The magnitude of the correction depends upon the sample shape and is zero for spherical samples, so that this disadvantage can be minimized by constructing special concentric spherical sample holders. The method is used with very reactive samples or with samples where lack of contamination is important.

Because the capillary holding the standard distorts the magnetic field around it, the field homogeneity in the annular outer part of the sample is destroyed. This can be restored by spinning, which is essential with this type of standardization. Distorted capillaries can even then degrade the resolution, and if there is any asymmetry in the annular region this will be averaged by spinning to give a field different in value from the true average, i.e. a small shift error will result. The method is nevertheless the only one suitable for measuring solvent shifts.

(c) Substitution

In this case the sample and standard are placed in separate tubes of the same size and are recorded separately in the order standard, sample, standard, so that if there is no lock available any field drift that occurs during the measurements can be allowed for. If the samples both contain a compound that can provide a lock (section 5.11), then their frequencies can be compared directly in two measurements, though with the same reservations as for the previous technique above. The locking system does, of course, allow standardization of a spectrum of one isotope species using another, either that of the lock or that of a third via the lock substance. In fact, today, many spectra are recorded for relatively dilute samples in deuterated lock solvents so that all are in effect referred to the same internal secondary standards. In the majority of cases, therefore, solvent effects and susceptibility changes are ignored unless particularly accurate comparative work is required. Indeed, this is the method commonly used for referring all but proton and carbon spectra, where TMS is usually added to the sample.

Spinning sidebands are often seen in spectra flanking the more intense lines and spaced equally to either side of them. They are only seen when the sample is spinning and their spacing from the central line depends on the speed of spinning. They arise because the magnetic field homogeneity is not perfect so that an individual nucleus, as it moves in the circular path impressed by the spinning, experiences a regularly fluctuating field. Their resonance width is reduced but their frequency is modulated, and a series of sidebands are produced separated by the spinning frequency. These become weaker if the spinning speed is increased, though the speed is limited by the tendency for the sample to be thrown up the wall of the tube and spoil the homogeneity, unless a plug is pushed down the tube to confine the sample.

2.5.2   Solvent effects

The solvent shift effects mentioned under internal standardization (section 2.5.1(a)), while a nuisance to those interested simply in structure determination, are of interest in their own right, since they tell us something about the weak interactions that occur between solvent and solutes. The effect is particularly large for aromatic solvents or where specific interactions occur. The chemical shifts of a number of substances relative to TMS in a series of solvents are given in Table 2.1.

The variation in δ between solvents, of course, contains contributions from the solvent effect on both solute and standard. Table 2.1 nevertheless is useful in indicating the existence of certain interactions involving the solutes. Thus the high-field shifts obtained in the aromatic solvent benzene and for all solutes but chloroform in the aromatic solvent pyridine are obvious. Even the relatively inert cyclohexane is shifted upfield by 0.05 ppm. This arises because solutes tend to spend a larger amount of time face on to the disc-shaped aromatic molecules and so on average are shifted upfield by the ring current anisotropy. The magnitude of the effect depends upon molecular shape and is also increased if there is any tendency for polar groups in the molecule to interact with the aromatic π electrons.

In the case of complex solutes, each type of proton in the molecule suffers

Table 2.1   *Internal chemical shifts δ of solutes in different solvents*

Solute	Solvent				
	$CDCl_3$	$(CD_3)_2SO$	Pyridine	Benzene	CF_3COOH
Acetone, $(CH_3)_2CO$	2.17	2.12	2.00	1.62	2.41
Chloroform, $CHCl_3$	7.27	8.35	8.41	6.41	7.25
Dimethyl-sulphoxide, $(CH_3)_2SO$	2.62	2.52	2.49	1.91	2.98
Cyclohexane, C_6H_{12}	1.43	1.42	1.38	1.40	1.47

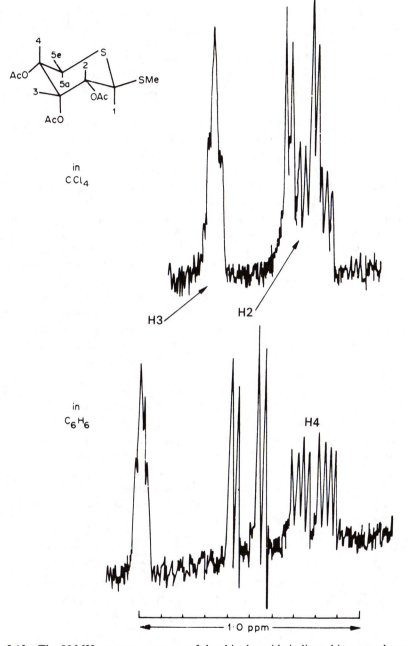

Fig. 2.13   The 90 MHz proton spectrum of the thioglycoside indicated in two solvents, tetrachloromethane (carbon tetrachloride) and benzene. Note the marked improvement that occurs in the latter solvent.

a solvent shift, but because the proximity of each to solvent depends upon the shape of the molecule, each suffers a different solvent shift. For this reason a complex solute may have quite different spectra in, for instance, chloroform and benzene, and a change from one solvent to the other may remove some degeneracy or avoid the overlapping of signals that would otherwise be difficult to disentangle. Figure 2.13 shows the quite large changes that can occur. This technique is often known by the acronym ASIS, or assisted solvent-induced shifts.

Chloroform as a solute suffers considerable solvent shifts. The pure liquid is self-associated by hydrogen bonding, but upon progressive dilution in an inert solvent the proportion of hydrogen-bonded molecules is reduced and its resonance is shifted 0.29 ppm upfield. The shifts noted in Table 2.1 are in excess of this and we must consider the existence of several other types of interaction. Thus specific interactions with the Lewis bases dimethylsulphoxide and pyridine result in low-field shifts. In the case of pyridine this implies a preference for an edge on approach to the aromatic ring and therefore some ring current de-shielding. In benzene, on the other hand, hydrogen bonding is reduced, and there is probably face-wise interaction of the chloroform with the benzene π electrons. Both processes tend to increase the screening, so that the chloroform is shifted strongly upfield. In addition, the choloform molecule is polar and its dipole electric field will polarize the surrounding solvent by an amount related to the sovent dielectric constant ε. This induced charge gives rise to an electric field, which is called the reaction field and which will also produce chemical shifts of the chloroform solute. Thus some of the variation observed in Table 2.1 will originate from differences in solvent dielectric constant.

Two further contributions to solvent shifts are also usually considered. One arises from the van der Waals interactions and is responsible for vapour–liquid shifts of 0.1 to 0.5 ppm. The other arises in the case of external standardization and is due to bulk diamagnetic susceptibility differences between solvents. These susceptibility shifts can be comparable in magnitude with those due to the other effects and so must be considered in any interpretation of solvent shifts, but they do not, of course, arise from any chemical interaction.

The various contributions to the solvent shift δ_S can be summarized by a live-term equation:

$$\delta_S = \delta_B + \delta_A + \delta_E + \delta_H + \delta_W$$

where δ_B is the bulk susceptibility contribution, δ_A is the anisotropy contribution, δ_E is the reaction field contribution, δ_H is the contribution of hydrogen bonding and specific interactions, and δ_W is the van der Waals contribution.

Questions

2.1. Which of the two chemically different types of protons in $CH_2ClCHCl_2$ resonate at higher frequency?

2.2. A proton spectrometer operating at 100 MHz was used to measure the frequency separation of the resonances of chloroform, $CHCl_3$, and TMS, which was found to be 730 Hz, the $CHCl_3$ being to high frequency. What is the chemical shift of chloroform on the δ scale? What would the frequency separation and chemical shift be if the sample were measured in a spectrometer operating at 350 MHz?

2.3. The 100 MHz spectrometer above is used to produce the proton spectrum of a complex molecule. Would the resolution be better or worse if the ^2H spectrum of the fully deuterated form of the same molecule were obtained? Assume that the linewidths of the ^1H and ^2H resonances are the same. The frequency used for ^2H spectra is 15.35 MHz. Assume also that there is no primary isotope effect.

3 Internuclear spin–spin coupling

3.1 The mutual effects of nuclear magnets on resonance positions

The Brownian motion in liquid samples averages the through-space effect of nuclear magnets to zero. However, in solutions of $POCl_2F$, for example, the phosphorus nucleus gives two resonances whose separation does not depend upon the magnetic field strength. (The chlorine nuclei ($I = 3/2$) have no effect. This is explained in Chapter 4.) This suggests that the two resonances correspond to the two spin orientations of the fluorine nucleus and that the nuclei *are* able to sense one another's magnetic fields. Theoretical considerations indicate that the interaction occurs via the bonding electrons. The contact between one nucleus and its *s* electrons perturbs the electronic orbitals around the atom and so carries information about the nuclear energy to other nearby nuclei in the molecule and perturbs their nuclear frequency. The effect is mutual and in the molecule mentioned above both the fluorine ($I = 1/2$) and the phosphorus ($I = 1/2$) resonances are split into doublets of equal hertz separation. The magnitude of the effect for a particular pair of nuclei depends on the following factors:

1. The nature of the bonding system, i.e. upon the number and bond order of the bonds intervening between the nuclei and upon the angles between the bonds. The interaction is not usually observed over more than five or six bonds and tends to be attenuated as the number of bonds increases, though many cases are known where coupling over two bonds is less than coupling over three'bonds.
2. The magnetic moments of the two nuclei and is directly proportional to the product $\gamma_A\gamma_B$ where γ_A and γ_B are the magnetogyric ratios of the interacting nuclei.
3. The valence *s* electron density at the nucleus and therefore upon the *s* character of the bonding orbitals. This factor also means that the interaction increases periodically as the atomic number of either or both nuclei is increased in the same way as does the chemical shift range.

The magnitude of the coupling interaction is measured in hertz (Hz) since it is the same at all magnetic fields. It is called the coupling constant and is given the symbol J; its magnitude is very variable and values have been reported from 0.05 Hz up to several thousand hertz. The value of J gives information about the bonding system but this is obscured by the contribution of γ_A and γ_B to J. For this reason correlations between the bonding system and spin–spin coupling often use the reduced coupling constant K, which is equal to $4\pi^2 J/\hbar\gamma_A\gamma_B$.

It is important to understand that coupling constants can be either positive or negative and that the frequency of one nucleus may be either increased or decreased by a particular orientation of a coupled nucleus, the sign depending upon the bonding system and upon the sign of the product $\gamma_A\gamma_B$.

Considerable data are available upon the magnitudes of interproton spin coupling constants from the mass of data accumulated for organic compounds. Interproton coupling is usually (though not always) largest between geminal protons (H–C–H), and depends upon the angle between the two carbon–hydrogen bonds. J_{gem} is typically 12 Hz in saturated systems. J falls rapidly as the number of intervening bonds is increased, being 7 to 8 Hz for vicinal protons (H–C–C–H) and near zero across four or more single bonds. The same rules apply if oxygen or nitrogen forms part of the coupling path, and methoxy protons (H_3COCHR_2) do not usually show resolvable coupling to the rest of the molecule, though alcoholic or amino protons may do so to vicinal protons in e.g. $HOCH_3$. On the other hand, coupling may be enhanced if there is an unsaturated bond in the coupling path, due to a σ–π configuration interaction, and may be resolved over up to as many as nine bonds, e.g. $^9J(H\!-\!H) = 0.4\,Hz$ between the hydrocarbon protons in $H_3C(C\!\equiv\!C)_3CH_2OH$. In saturated molecules, a planar zig-zag configuration of the bonds may also lead to resolvable coupling over four or five single bonds. Note the use of the superscript 9 in the acetylene example to indicate the number of bonds over which the interaction occurs.

Karplus has calculated the values of the vicinal interproton coupling constants and shown that these depend upon the dihedral angle ϕ between the carbon–hydrogen bonds (Fig. 3.1). Unfortunately, the magnitude of coupling is also influenced by such factors as the nature of the other substituents on the two carbon atoms, their electronegativity, their orientation, the hybridization at the carbon, the bond angles other than the dihedral angle and the bond lengths. For this reason two curves are shown, which give an idea of the range of coupling constants that may be encountered. If the geometry of the molecule is fixed, then the value of the coupling constant enables an estimate to be made of the value of ϕ. If the geometry is not fixed, then the coupling constant is an average over the possible values of ϕ. If free rotation is possible, then the average is over the whole of the curves illustrated. Thus in fluoroethane (ethyl fluoride) and ethyllithium the vicinal coupling is 6.9 and 8.4 Hz respectively, the value increasing with increasing substituent electronegativity. If the geometry is static, then the values of the coupling constants may show quite clearly what the conformation of the molecule studied has to be. Thus in Fig. 3.2, part of the proton spectrum of a thioglycoside is given in which various doublet splittings due to spin–spin coupling are evident. The resonance of H2 is split into two doublets with a small J of 2.6 Hz and a large one of 10.6 Hz. H1 is easily

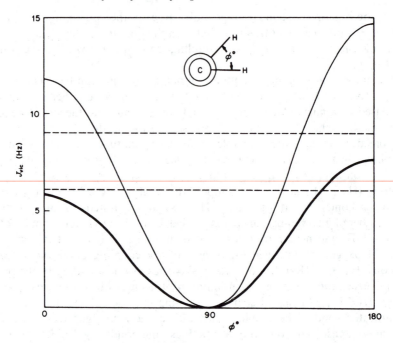

Fig. 3.1   Karplus curves relating the dihedral angle ϕ in a HC–CH fragment and the vicinal proton–proton coupling constant. The inset shows a view along the carbon–carbon bond. Two curves are shown relating to differently substituted fragments, and are differentiated by heaviness of line. The dashed lines show the typical range of values obtained when a group can rotate freely, giving rise to an averaged J_{vic}. (After Jackman and Sternhell, *NMR Spectroscopy in Organic Chemistry*, Pergamon Press 1969, with permission.)

identifiable since it is the only proton with a single doublet splitting, and it is the large coupling J(H1–H2) that is due to this axial–axial pair with a dihedral angle of near 180° and large predicted J. The other lesser interaction is J(H2–H3), H2 and H3 being an axial–equatorial pair with dihedral angle of near 90° and so small predicted J. Evidently the splittings in the H3 resonance are all small and both neighbouring hydrogen atoms must form axial–equatorial pairs.

This vicinal ^3J dependence on dihedral or torsion angle seems to be quite general and Karplus-type curves have been established for the coupling paths ^{13}C–C–C–^{1}H, ^{31}P–C–C–^{31}P and ^{13}C–C–C–^{31}P and are also likely for coupling between ^{77}Se or ^{125}Te and ^{1}H or ^{13}C or for the more exotic systems such as $^3J(^{199}$Hg–^{13}C) in alkylmercury compounds.

3.2   The appearance of multiplets arising from spin–spin coupling

The appearance of these multiplets is very characteristic and contains much information additional to that gained from chemical shift data. We have already

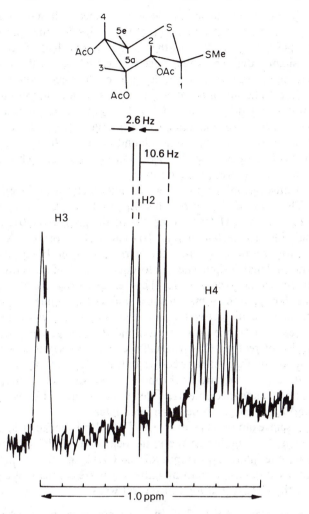

Fig. 3.2   Part of the 90 MHz proton spectrum of the thioglycoside whose structure is shown in the inset. The numbers represent hydrogen atoms attached to the ring. The resonances of H3, H2 and H4 are shown and are split by spin–spin coupling between them and H1 and H5 equatorial and H5 axial.

seen in Fig. 3.2 that coupling to a single proton or protons with different coupling constants gives a series of lines of approximately equal height. Where coupling is to groups of nuclei or between different nuclear species, then the patterns can be very complex, and unravelling them can tell us much about the structure of fragments of a molecule. The simplest case to consider is the effect that a single chemically unique nucleus of $I = 1/2$ has no other nuclei in the molecule that are sufficiently closely bonded. In half the molecules in the sample, the spin of our nucleus N will be oriented in the same direction as the field and all the other nuclei in these molecules will have corresponding resonance positions. In the remainder of the molecules, the spin of N will be opposed to the field and all the other nuclei in this half of the sample will resonate at slightly different frequencies to their fellows in the first half. Thus when observing the sample as a whole, each of the nuclei coupled to N gives rise to two lines. The line intensities appear equal since the populations of N in its two states only differ by about 6 in 10^5, which is not detectable. We say that N splits the other resonances into 1:1 doublets. Because the z component of magnetization of N has the same magnitude in both spin states, the lines are equally displaced from the chemical shift positions of each nucleus, which are therefore at the centres of the doublets.

Let us in illustration consider the molecule $CHCl_2CH_2Cl$ and its proton resonance. This contains two sorts of hydrogen, with the $CHCl_2$ proton resonating to low field of the CH_2Cl protons due to the greater electric field effect of two geminal chlorine–carbon bonds. The two CH_2Cl protons have the same frequency since rotation around the carbon–carbon bond averages their environments and makes them chemically equivalent. It also averages the vicinal coupling to the single proton to be the average of the Karplus curve. These two protons thus have the same chemical shift and the same coupling constant to all other magnetically active nuclei in the sample; a trivial condition in this case since there is only one other proton and the spectra are indifferent to the chlorine nuclei for reasons that we shall see later. We say that the CH_2 nuclei are isochronous and magnetically equivalent. They are, of course, coupled quite strongly, but because they are isochronous they resonate as if they were a unit and give a singlet resonance unless coupled to other nuclei. This is a consequence of the second-order effects to be considered later.

In the present example, however, the CH_2Cl resonance is split into a 1:1 doublet because of coupling to the non-isochronous $CHCl_2$ proton. Equally, since the coupling interaction is mutual, the $CHCl_2$ proton is split by the two CH_2Cl protons, though the splitting pattern is more complex. We can discover the shape of the $CHCl_2$ multiplet in several ways.

1. By an arrow diagram (Fig. 3.3). In some molecules both the CH_2Cl spins will oppose the field, in others both spins will lie with the field, while in the remainder they will be oriented in opposite directions. The $CHCl_2$ protons in the sample can each experience one of three different perturbations and their resonance will be split into a triplet. Since the CH_2Cl spins can be paired in opposition in two different ways, there will be twice as many molecules with them in this state as there are with them in each of the other two. The $CHCl_2$ resonance will therefore appear as a 1:2:1 triplet with the spacing between the lines the same as that of the CH_2Cl 1:1 doublet. An actual spectrum is shown in Fig. 3.4, where it should be noted that the intensities are not exactly as predicted by the simple first-order theory but are perturbed by second-order effects. Note also that the total multiplet intensity is

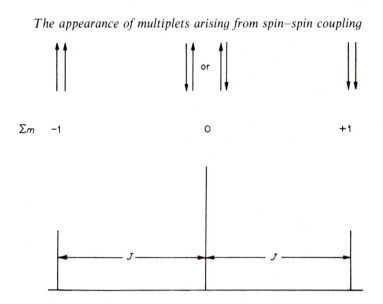

Fig. 3.3   A stick diagram demonstrating the splitting due to two spin-1/2 nuclei. When $\Sigma m = 0$ there is no perturbation of the coupled resonance so that the centre line corresponds to the chemical shift position. This holds for all multiplets with an odd number of lines. The spacing J between the lines corresponds to a change in Σm of unity.

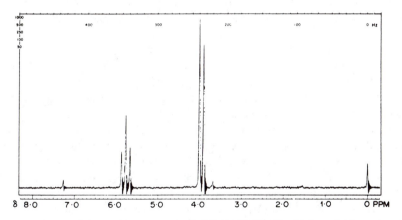

Fig. 3.4   The 60 MHz proton spectrum of $CH_2ClCHCl_2$. The highest-field resonance is due to the protons of TMS. The compound was dissolved in deuteriochloroform ($\sim 7\%$ solution) and the 0.5% or so of protons in the solvent appears at 7.3 ppm. The resonances are asymmetric with a ringing pattern to high field of each. This indicates that they were obtained in the CW mode. (Reproduced by permission of Varian International AG.)

proportional to the number of protons giving rise to each multiplet. The coupling pattern thus counts precisely the number of protons in interaction and the intensity enables us to relate different, non-coupled groups of lines in the spectrum.

2. We can also work out the multiplicity by considering the possible values of the total magnetic quantum number Σm of the two CH_2Cl protons. We have $I = 1/2$, and m can be $\pm 1/2$. Therefore for two protons (Fig. 3.3)

$$\Sigma m = +1/2 + 1/2 = +1$$

or

$$\Sigma m = \begin{cases} +1/2 - 1/2 = 0 \\ -1/2 + 1/2 = 0 \end{cases}$$

or

$$\Sigma m = -1/2 - 1/2 = -1$$

Therefore we have three lines.

Methods 1 and 2 are equivalent, but method 2 is particularly useful when considering multiplets due to nuclei with $I > 1/2$ where arrow diagrams become rather difficult to write down clearly. Note that when $\Sigma m = 0$ there is no perturbation of the chemical shift of the coupled group so that the centre of the spin multiplet corresponds to the chemical shift of the group.

Next let us consider the very commonly encountered pattern given by the ethyl group CH_3CH_2-. The isochronous pair of CH_2 protons are usually found to low field of the CH_3 protons and are spin-coupled to them. The CH_3 protons therefore resonate as a 1:2:1 triplet. The splitting of the CH_2X resonance caused by the CH_3 group can be found from Fig. 3.5, and is a 1:3:3:1 quartet. A typical ethyl group spectrum is shown in Fig. 3.6.

We have done enough now to formulate a simple rule for splitting due to groups of spin-1/2 nuclei. Thus the number of lines due to coupling to n equivalent spin-1/2 nuclei is $n + 1$. The intensities of the lines are given by the binomial coefficients of $(a + 1)^n$ or by Pascal's triangle, which can be built up as required. This is shown in Fig. 3.7. A new line of the triangle is started by writing a 1 under and to the left of the 1 in the previous line and then continued by adding adjacent figures from the old line in pairs and writing down the sum as shown. The multiplicity enables us to count the number of spin-1/2 nuclei in a group and the intensity rule enables us to check our assignment in complex cases where doubt may exist, since the outer components of resonances coupled to large groups of nuclei (e.g. the CH of $(CH_3)_2CH-$) may be too weak to observe in a given spectrum.

Coupling to nuclei with $I > 1/2$ leads to different relative intensities and multiplicities. In the case of a single nucleus the total number of spin states is equal to $2I + 1$ and this equals the multiplicity. If $I = 1/2$ we get two lines,

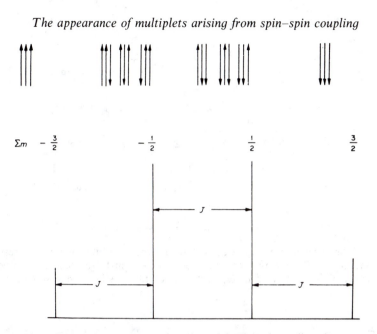

Fig. 3.5 The splitting due to three spin-1/2 nuclei. There is no line that corresponds to $\Sigma m = 0$. The multiplet is, however, arranged symmetrically about the $\Sigma m = 0$ position, so that the centre of the multiplet corresponds to the chemical shift position. This rule holds for all multiplets with an even number of lines.

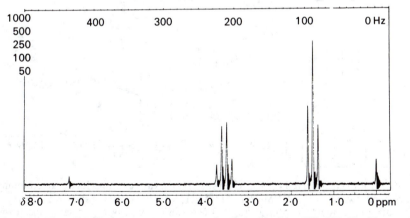

Fig. 3.6 The 60 MHz proton spectrum of ethyl chloride CH_3CH_2Cl. Also a CW spectrum. (Reproduced by permission of Varian International AG.)

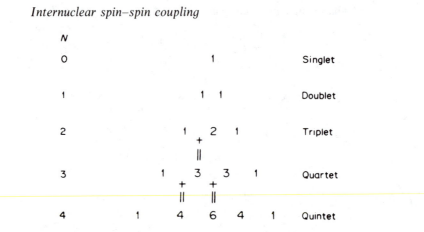

N						
0			1			Singlet
1			1 1			Doublet
2			1 2 1			Triplet
3		1 3 3 1				Quartet
4	1	4	6	4	1	Quintet

Fig. 3.7   Pascals triangle can be used to estimate the intensities of the lines resulting from coupling to different numbers, N, of equivalent spin-1/2 nuclei. The numbers in each line are obtained by adding adjacent pairs of numbers in the line above.

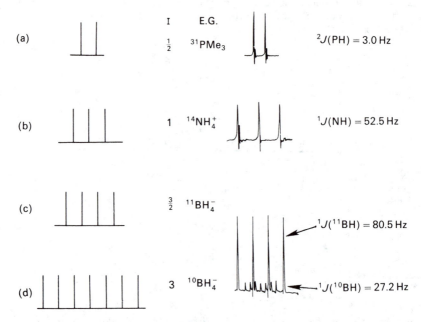

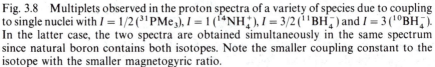

Fig. 3.8   Multiplets observed in the proton spectra of a variety of species due to coupling to single nuclei with $I = 1/2\,(^{31}PMe_3)$, $I = 1\,(^{14}NH_4^+)$, $I = 3/2\,(^{11}BH_4^-)$ and $I = 3\,(^{10}BH_4^-)$. In the latter case, the two spectra are obtained simultaneously in the same spectrum since natural boron contains both isotopes. Note the smaller coupling constant to the isotope with the smaller magnetogyric ratio.

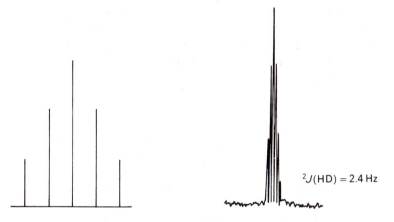

$$^2J(\text{HD}) = 2.4\,\text{Hz}$$

Fig. 3.9   Multiplet pattern obtained for a proton coupled to two deuterium nuclei with $I = 1$ in the minor component in $(^2H_6)$ acetone (deuterioacetone) containing 0.5% hydrogen. The value of $^2J(^1H-^1H)$ can be calculated from this spectrum. This coupling exists but cannot be measured in the all-hydrogen form.

$I = 1$ gives three lines, $I = 3/2$ gives four lines, and so on. The spin populations of each state are virtually equal and so the lines are all of equal intensity and of equal spacing (Fig. 3.8).

Splitting due to multiple combinations of $I > 1/2$ nuclei is much less common, but a few examples have been recorded. Figure 3.9 illustrates the pattern commonly encountered when obtaining ^1H spectra in the solvent $(^2H_6)$ acetone (deuterioacetone, $(CD_3)_2CO$). In practice, there will always be a small amount of hydrogen present in these molecules and some of the methyl groups will be CD_2H- groups. The proton sees two equivalent deuterons with $I = 1$. The maximum total Σm is $1 + 1 = 2$. Thus $\Delta m = \pm 1$ and there are therefore five spin states and the proton resonance will be a quintet. In order to determine the line intensities, we have to determine the number of ways each value of Σm can be obtained. This is shown in Table 3.1 and indicates relative intensities of 1:2:3:2:1, a distribution that differs from the simple binomial.

Table 3.1   Line intensities for coupling to two nuclei with $I = 1$

Σm	Possible spin combinations	Number of spin combinations
2	$(+1, +1)$	1
1	$(+1, 0), (0, +1)$	2
0	$(+1, -1), (0, 0), (-1, +1)$	3
-1	$(-1, 0), (0, -1)$	2
-2	$(-1, -1)$	1

Thus the rule given for spin-1/2 nuclei can be generalized to include groups of nuclei of any given I, the number of lines observed for coupling to n equivalent nuclei of spin I being $2nI + 1$. Obtaining the relative intensities is, however, tedious and it is better to use a Pascal's triangle type of construction as shown below:

```
                    I = 1                                    I = 3/2
 n
 0              1                                        1
 1           1  1  1                              1   1   1   1
 2         1  2  3  2  1                        1   2   3   4   3   2   1
 3       1 [3  6  7] 6  3  1                  1   3 [6  10  12  12] 10  6   3   1
 4     1  4 10 16 19 16 10  4  1    1   4   10  20  31  40  44  40  31  20  10   4   1
```

In the case of $I = 1$, the numbers in the triangle are placed immediately under the previous ones, since there is always a line in the centre of the multiplet. For $I = 3/2$ (and all half-integral spins), the numbers are staggered, since only if n is even is there a line at the centre of the multiplet. The triangles are constructed by moving a box, which can enclose $2I + 1$ numbers of a line, along the line, enclosing first the 1, then two numbers, then three, then four, and so on. The numbers enclosed by the box are added to give a number for the next line. Thus for $I = 3/2$ the box can enclose four numbers, and if we sweep through the line for $n = 1$, we enclose first a 1, then $1 + 1 = 2$, then $1 + 1 + 1 = 3$, then $1 + 1 + 1 + 1 = 4$ and reducing again to $3, 2, 1$. The construction given above for spin-1/2 nuclei is the first example of this series.

More complex coupling situations also arise where a nucleus may be coupled simultaneously to chemically different groups of nuclei of the same or of different isotopes or species. The patterns are found by building up spectra, introducing the interactions with each group of nuclei one at a time. Thus Fig. 3.10 shows how a group M coupled to two chemically different spin-1/2 nuclei A and X is first split by $J(A–M)$ into a doublet, and shows that each doublet line is further split by $J(M–X)$. If $J(A–M) = J(A–X)$, a 1:2:1 triplet is obtained; but if $J(A–M) \neq J(A–X)$, then a doublet of doublets with all lines of equal amplitude arises. This can be distinguished from a ^{11}B coupling because the line separations are irregular, and of course the preparative chemist is usually aware whether or not there should be boron in the compound.

When analysing such multiplets, it always has to be borne in mind that overlap of lines may occur so that fewer than the theoretical number of lines are observed and the intensities are unusual. Such a case is illustrated in Fig. 3.11, which shows the ^9Be spectrum of the complex $(\eta^5\text{-}C_5H_5)BeBH_4$. The ^9Be resonance is split by coupling to the nuclei of the BH_4 ligand only, with $J(^9Be\text{-}^1H) = 10.2\,Hz$ and $J(^9Be\text{-}^{11}B) = 3.6\,Hz$. The first coupling causes splitting to a 1:4:6:4:1 quintet, each line of which is further split into a 1:1:1:1 quartet by the ^{11}B. So 20 lines are expected, but because the ratio of the coupling constants is almost integral (it is 2.8) some lines overlap and only 16 are observed. The overlap is not exact but the close pairs of lines are not resolvable because of the appreciable natural linewidth of the ^9Be resonance.

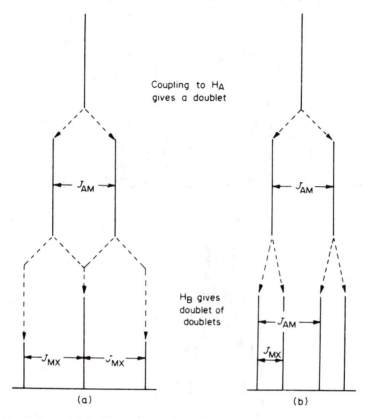

Coupling to H_A
gives a doublet

H_B gives
doublet of
doublets

(a)   (b)

Fig. 3.10   Splitting of the H_M protons of Z_2CH_A–$C(H_M)_2$–CH_XY_2 due to coupling to H_A and H_X: (a) $J(A-M) = J(X-M)$, the centre lines overlap and the multiplet is a 1:2:1 triplet just as if H_M were coupled to a CH_2 group; (b) $J(A-M) \neq J(A-M)$ and we get a doublet from which both $J(A-M)$ and $J(X-M)$ can be measured.

The same technique can be used to predict the shapes of multiplets due to several equivalent nuclei of spin $I > 1/2$, i.e. introducing the splitting due to one nucleus several times as shown in Fig. 3.12 for two spin-3/2 nuclei. Comparison with the Pascal's triangle diagrams will show that the two methods are exactly equivalent.

3.3   Spin–spin coupling satellites

A glance at the table of nuclear properties (Table 1.1) will show that certain elements have as principal isotope a magnetically non-active species, but that they have also a more or less small proportion of magnetically active species,

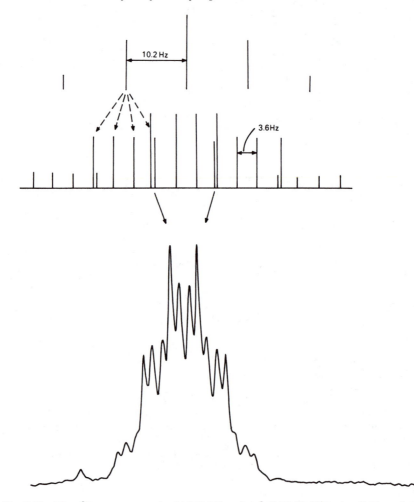

Fig. 3.11 The ^{9}Be spectrum at 14.06 MHz of $(\eta^5\text{-}C_5H_5)BeBH_4$ in C_6F_6 solution. Coupling is observed to both ^{11}B ($I = 3/2$) ($J = 3.6$ Hz) and ^{1}H ($I = 1/2$) ($J = 10.2$ Hz) of the BH_4 group. Some lines overlap and the multiplicity is less than the maximum given by the basic rules. (From Gaines *et al.* (1981) *J. Magn. Reson.*, **44**, with permission)

some with $I = 1/2$. Examples are platinum with 33.8% of the active nucleus ^{195}Pt and carbon with just 1.1% of the active ^{13}C. The spectra of other nuclei in compounds of these elements will thus arise from differentiable molecular species: those with the non-active isotope, which will give rise to intense patterns depending upon the other nuclei present, and those with the active isotope, which will give a weaker sub-spectrum in which spin–spin coupling will be seen arising from interaction with this less abundant isotope. These weak doublets are centred approximately around the corresponding lines in the spectrum of

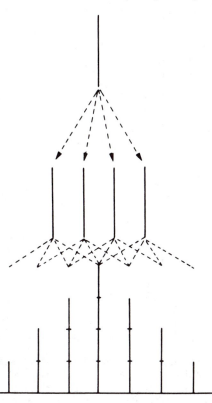

Fig. 3.12   An illustration showing how by successively introducing the quartet splitting due to coupling to two equivalent nuclei with $I = 3/2$ the correct 1:2:3:4:3:2:1 pattern is obtained.

the main species and are thus called satellites. They are usually particularly obvious in ^1H or ^{31}P spectra of platinum compounds, since their intensity is about a quarter of that of the central line of the major species. We will give two examples of compounds containing one and two platinum atoms.

First, we will consider the proton spectrum of the mononuclear square planar complex *trans*-MeBrPt(PMe$_2$R)$_2$, where R is a 2,4-dimethoxyphenyl group. The methyl groups attached to Pt and to phosphorus are both coupled to ^{195}Pt, and their spectra, in the high-field region, are shown in Fig. 3.13, with the effect of the ^{31}P atoms removed by a process of double irradiation, which we will discuss in detail in a later chapter. There are two major methyl resonances marked by a star and which originate from the complexes with magnetically inactive Pt, and each is flanked by two lines of quarter intensity, which are the doublets due to coupling to one spin-1/2 nucleus, ^{195}Pt. The shorter-distance coupling path gives an appreciably stronger coupling than the longer-distance

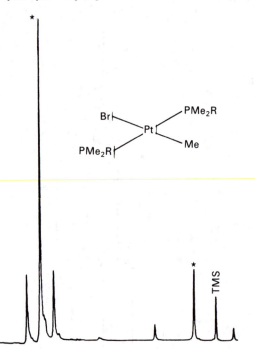

Fig. 3.13   The ¹H spectrum of the complex shown, with the effect of the ³¹P spin-1/2 nuclei removed by a double irradiation technique. The P–Me and Pt–Me resonances are 1:4:1 triplets, the weaker lines in each triplet being spin coupling satellites due to protons coupled to the minor platinum isotope ¹⁹⁵Pt. (Spectrum supplied by Professor B. L. Shaw.)

one via P. Evidently, the values of the coupling constants give structural information and the existence of the 1:4:1 pattern is good proof of the presence of ¹⁹⁵Pt in the complex. If we now consider what sort of molecular populations we might expect if we have a complex that contains two platinum atoms, we will realize that we should have three sub-spectra, one each from those complexes which contain no ¹⁹⁵Pt, those which contain one and the smaller number which contain two. If the complex is symmetric, such as

$$\text{Ph}_3\text{P.Pt(CO)Pt(CO).PPh}_3$$
with an S bridge

then we get three interlaced spectra as shown in Fig. 3.14, which gives the ³¹P spectrum with the couplings to the protons now suppressed by double irradiation. This complex contains 43.8% of its molecules with no ¹⁹⁵Pt in which the phosphorus atoms are entirely equivalent and give a central, strong singlet. Some 44.8% of the molecules contain one ¹⁹⁵Pt atom and this is coupled unequally to the two phosphorus nuclei with the one-bond coupling constant

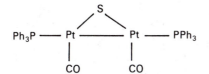

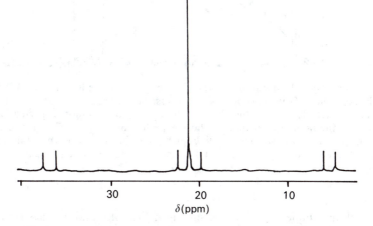

Fig. 3.14   The ^{31}P spectrum of the binuclear platinum cluster complex shown, with the effect of the coupling to the protons removed by a double irradiation technique. The spectrum consists of three sub-spectra arising from molecules with none, with one and with two ^{195}Pt atoms in the cluster. (From Evans *et al.* (1987) *J. Chem. Soc., Dalton Trans.*, 1889, with permission.)

being 3161 Hz and the two-bond coupling being 122 Hz. This makes the two phosphorus atoms non-equivalent and they also can couple with a coupling value of 149 Hz. The resulting sub-spectrum contains two pairs of doublets, one widely spaced due to the ^{31}P directly bonded to ^{195}Pt and one with two lines overlapping the central resonance due to the other ^{31}P nucleus. The remaining 11.4% of the molecules contain two ^{195}Pt atoms and give a weak, poorly resolved spectrum, which can just be discerned near the central and outermost lines. This sub-spectrum is a second-order spectrum of a type to be discussed in detail below. The spectra become much more complex for bigger clusters and are of considerable use in studying the structures of this interesting class of compound.

Another nucleus that provides interesting examples of spin coupling satellites is ^{183}W, which has a natural abundance of 14.28%. The ^{19}F spectrum of the

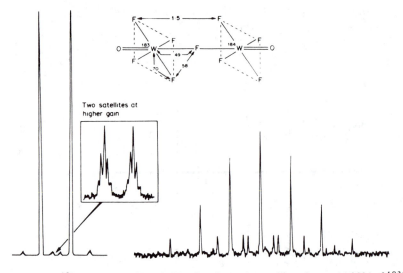

Fig. 3.15   The ^{19}F spectrum of $W_2O_2F_9^-$ showing spin satellites due to 14.28% of ^{183}W. The outer lines of the nonet are lost in the baseline noise and the student should confirm that the intensity ratios of the observed lines correspond to those expected for the inner seven of the nonet rather than to those expected for a septet. The arrows around the formula indicate the various coupling interactions in hertz. The single fluorine nonet is recorded at higher gain. (After McFarlane *et al.* (1971) *J. Chem. Soc. A*, 948, with permission.)

binuclear complex $W_2O_2F_9^-$ is shown in Fig. 3.15, and consists principally of a doublet of intensity 8 and a nonet of intensity 1. The nine fluorine atoms can thus be divided into an isochronous set of eight atoms and one unique atom. The main spectrum arises from those molecules in which both tungsten atoms are the magnetically inactive ^{184}W but satellite lines are also observed due to the molecules with one ^{183}W atom in their structure. In principle, lines should also exist due to those molecules with two ^{183}W atoms, but their proportion is low and their resonances are too weak to observe. Each of the lines of the intense doublet has two ^{183}W satellites, each of which is further split into a 1:4:6:4:1 quintet. This pattern must arise from coupling to four fluorine atoms. We can therefore conclude that we have four of the eight isochronous fluorine atoms associated with the ^{183}W atom and therefore split into a satellite doublet and then further coupled to the remaining four, which are equally associated with the ^{184}W. This provides considerable confirmatory evidence that the structure is OWF_4. F. WF_4O with a fluorine atom bridging the tungsten atoms.

Finally we must consider the effect that the 1.1% of naturally occurring ^{13}C has on the proton spectra of organic compounds, which contain principally the inactive ^{12}C. In fact, the spin satellites due to this minor isotopic component are not easy to observe among the intense ^{1}H–^{12}C resonances, but have proved to be of considerable use. In simple compounds such as acetone, $(CH_3)_2CO$, for instance, the proton resonance has two pairs of spin coupling satellites due to molecules with the ^{13}C in the methyl group ($^1J(^1H-^{13}C) = 126$ Hz) and to those with ^{13}C in the carbonyl group ($^2J(^1H-^{13}C) = 5.9$ Hz). Thus we can measure proton–carbon coupling constants, and with double-resonance techniques we will see later that we can discover correlations between the proton and carbon spectra of a molecule.

3.4   Second-order effects

The rules so far discussed apply to spectra of nuclei of the same species where the separation between multiplets in hertz (i.e. the chemical shift) is large compared to the value of the coupling constant between them, or to coupling between nuclei of different elements or isotopes where the differences in NMR frequency are invariably large. A nomenclature has been adopted for these cases in which the chemically non-equivalent sets of spins are labelled with letters from the alphabet, choosing letters that are well separated in the alphabetic sequence to signify large chemical shift separation. Thus $CH_2ClCHCl_2$ is an A_2X system, CH_3CH_2R is an A_3X_2 system, and CH_3CH_2F is an A_3M_2X system. Their spectra are called first-order.

We have already seen in the examples that the line intensities in proton spectra exhibiting interproton coupling often do not correspond exactly to those predicted by the first-order rules, and these distortions increase as the interproton chemical shift is reduced. The spectra are said to become second-order, and to signify this and the fact that the chemical shifts between the coupled nuclei are relatively small, the spins are labelled with letters close together in the alphabet. Thus, for example, two coupled protons resonating close together are given the letters A B, and an ethyl group in $(CH_3CH_2)_3Ga$, where the methyl and methylene protons resonate close together, is described as an A_3B_2 grouping. Mixed systems are also possible and a commonly encountered one is the three-spin A B X grouping where two nuclei resonate close together and a third is well shifted or is of a different nuclear species.

Second-order spectra arise when the frequency separation between multiplets due to different equivalent sets of nuclei is similar in magnitude to the coupling constant between them; under these circumstances the effects due to spin coupling and chemical shift have similar energy and become intermingled, leading to alterations in relative line intensities and in line positions. Because it is the ratio between the frequency separation and J that is important, chemical shifts are always expressed in hertz (Hz) and not in parts per million (ppm) when discussing second-order spectra. The hertz separation is obtained by multiplying the chemical shift δ by the spectrometer operating frequency and is written $v_0\delta$. The perturbation of the spectra from the first-order appearance is then a function of the ratio $v_0\delta/J$ and is different for spectrometers operating at different frequencies. If a high enough frequency is used, many second-order spectra approach their first-order limit in appearance and are then much more easily interpreted. This is one of the advantages of high-field instrumentation. We will describe first the simplest possible system consisting of two spin-1/2 nuclei.

3.4.1   The AB second-order system

We consider a system consisting of two isolated but mutually spin-coupled spin-1/2 nuclei with different chemical shifts. When the chemical shift between them is large, much larger than the coupling, then we see two doublets. A typical arrangement with $J = 10\,\text{Hz}$ and $v_0\delta = 200\,\text{Hz}$ is shown in Fig. 3.17a. If we reduce the chemical shift progressively to zero, we can imagine the two doublets approaching one another until they coincide. However, we know that an isolated pair of equivalent nuclei give rise to a singlet, and the problem is how can two doublets collapse to give a singlet. If the coupling constant remains at 10 Hz, why do we not get a doublet A_2 spectrum?

The behaviour of the multiplets can be predicted using a quantum-mechanical arrangement with $J = 10\,\text{Hz}$ and $v_0\delta = 200\,\text{Hz}$ is shown in Fig. 3.17a. If we reduce the chemical shift progressively to zero, we can imagine the two doublets approaching one another until they coincide. However, we know that an isolated pair There are four such spin states with the wavefunctions $\alpha\alpha$, $\alpha\beta$, $\beta\alpha$, and $\beta\beta$, where the first symbol in each pair refers to the state of the A nucleus and the second symbol to the state of the B nucleus. The energies of the states $\alpha\alpha$ and $\beta\beta$ can be calculated straightforwardly, but the two states $\alpha\beta$ and $\beta\alpha$ have the same total spin angular momentum and it is found that the quantum-mechanical equations can only be solved for two linear combinations of $\alpha\beta$ with $\beta\alpha$. This is described as a mixing of the states and means that none of the observed transitions corresponds to a pure A or a pure B transition. the form of the wavefunctions and the energy levels derived are shown in Table 3.2. C_1 and C_2 are constants, v_A and v_B are the A and B nuclear frequencies in the absence of coupling and $v_0\delta = |v_A - v_B|$. The transition energies are the differences between four pairs of energy states, 3–4, 2–4, 1–3 and 1–2, each transition involving a mixed energy level. Because of the mixing, the transition probabilities are no longer equal as in the first-order case and intensity is transferred from lines in

Table 3.2   *Wavefunctions and energy levels for AB second-order system*

State	Wavefunction	Energy level (Hz)
1	$\alpha\alpha$	$\frac{1}{2}(v_A + v_B) + \frac{1}{4}J$
2	$C_1(\alpha\beta) + C_2(\beta\alpha)$	$\frac{1}{2}[(v_0\delta)^2 + J^2]^{1/2} - \frac{1}{4}J$
3	$-C_2(\alpha\beta) + C_1(\beta\alpha)$	$-\frac{1}{2}[(v_0\delta)^2 + J^2]^{1/2} - \frac{1}{4}J$
4	$\beta\beta$	$-\frac{1}{2}(v_A + v_B) + \frac{1}{4}J$

Table 3.3 Transition energies relative to centre of multiplet and relative intensities for AB second-order system

Transition	Energy (Hz)	Relative intensity
a $3 \rightarrow 1$	$+\frac{1}{2}J + \frac{1}{2}[(v_0\delta)^2 + J^2]^{1/2}$	$1 - J/[(v_0\delta)^2 + J^2]^{1/2}$
b $4 \rightarrow 2$	$-\frac{1}{2}J + \frac{1}{2}[(v_0\delta)^2 + J^2]^{1/2}$	$1 + J/[(v_0\delta)^2 + J^2]^{1/2}$
c $2 \rightarrow 1$	$+\frac{1}{2}J - \frac{1}{2}[(v_0\delta)^2 + J^2]^{1/2}$	$1 + J/[(v_0\delta)^2 + J^2]^{1/2}$
d $4 \rightarrow 3$	$-\frac{1}{2}J - \frac{1}{2}[(v_0\delta)^2 + J^2]^{1/2}$	$1 - J/[(v_0\delta)^2 + J^2]^{1/2}$

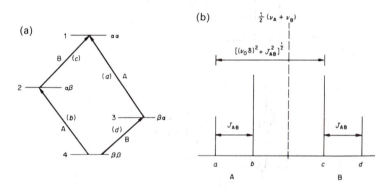

Fig. 3.16 The energy level diagram for a system of two spins related to the resulting AB quartet. The spectrum was calculated for $J = 10$ Hz, $v_0\delta = 28.2$ Hz, i.e. $v_0\delta/J = 2.82$.

the outer parts of the total multiplet into the central region. The transition energies relative to the centre of the multiplet, i.e. to the mean frequency $\frac{1}{2}(v_A + v_B)$, and the intensities are shown in Table 3.3.

There are thus four lines as in the AX spectrum but with perturbed intensities and resonance frequencies. The resulting spectral pattern and the corresponding energy level diagram are shown in Fig. 3.16. The energy levels are marked with the appropriate spin state and the transitions, which in the first-order case can be regarded as arising from transitions of the A or B nucleus, form opposite sides of the figure. The three line separations are a–b $= J$, c–d $= J$ and b–c $= [(v_0\delta)^2 + J^2]^{1/2} - |J|$. The separation a–c or b–d, which in the first-order case is the same as the separation between the doublet centres, and is therefore the chemical shift in hertz, $v_0\delta$, is now simply $[(v_0\delta)^2 + J^2]^{1/2}$ and is larger than the true chemical shift. In other words, though $v_0\delta$ is reduced to zero, the doublet centres never coincide and are separated by J Hz. The outer lines, however, have intensity zero at this point, while the inner lines are coincident, i.e. we predict a singlet spectrum as is observed (cf. Fig. 3.17). Thus arises our rule, 'isochronous coupled protons resonate as a unit'.

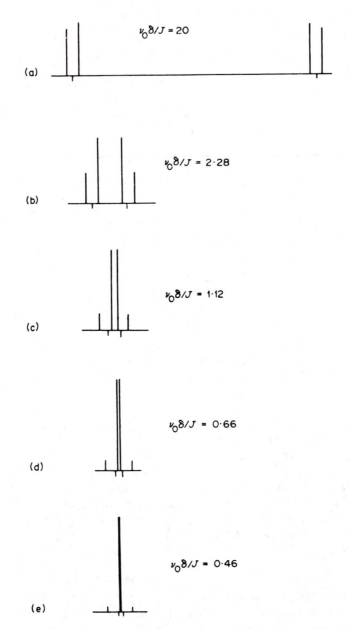

Fig. 3.17 AB quartets for several values of the ratio $v_0\delta/J$. The small markers under the baseline represent the true A and B chemical shifts.

We can calculate some simple rules for analysing an AB spectrum:

1. The spectrum contains two intervals equal to J, a–b and c–d.
2. The true AB chemical shift $v_0\delta$ is found as

$$(a–d)(b–c) = \{[(v_0\delta)^2 + J^2]^{1/2} + J\}\{[(v_0\delta)^2 + J^2]^{1/2} - J^2\}$$
$$= (v_0\delta)^2 + J^2 - J^2$$
$$= (v_0\delta)^2$$

so

$$v_0\delta = [(a–d)(b–c)]^{1/2}$$

where a–d is the separation between the outermost lines and b–c is the separation between the innermost pair of lines.
3. The assignment can be checked against the intensity ratios of the larger and smaller lines. The intensity ratio, stronger/weaker, is

$$\{1 + J/[(v_0\delta)^2 + J^2]^{1/2}\}/\{1 - J[(v_0\delta)^2 + J^2]^{1/2}\}$$

which gives

$$\{[(v_0\delta)^2 + J^2]^{1/2} + J\}/\{[(v_0\delta)^2 + J^2]^{1/2} - J\}$$

the ratio of the line separations (a–d)/(b–c). Note that changing the sign of J does not alter the pattern.

Figure 3.17 shows the form of the AB quartet for several values of $v_0\delta/J$. The true chemical shift positions are marked below the baseline. It is important to remember that if a multiplet shows signs of being highly second-order then both intensities *and* resonance positions are perturbed from their first-order values. A spacing corresponding to J_{AB} remains in AB-type spectra since only one coupling interaction exists, but in more complex systems the spacings are combinations of coupling constants. On the other hand, if the intensity perturbation is only slight (Fig. 3.17(a)) then the line positions are not detectably perturbed. Three examples of actual spectra that contain an AB multiplet are given in Fig. 3.18.

The spectra illustrated in Fig. 3.18 were obtained at relatively low frequency (60 MHz), and on modern instrumentation all the AB quartets could be converted to the simple first-order form by using spectrometer frequencies of 200, 400 or even 500 MHz giving up to 8.3 times increased ratio $v_0\delta/J$. Spectrum (c) of ascaridole, however, has some interesting additional features. The doublet marked A is due to coupling of two methyl groups to a single proton whose resonance, a septet, is lost among other signals in the region 1.5–2.0 ppm. This doublet also shows second-order perturbation of intensities, but, of course, less than for quartet CD since the chemical shift is larger. The other signals near 2 ppm arise from the $CH_2 \cdot CH_2$ ring fragment and are also second-order because rotational averaging of the proton positions, and so of chemical shift and

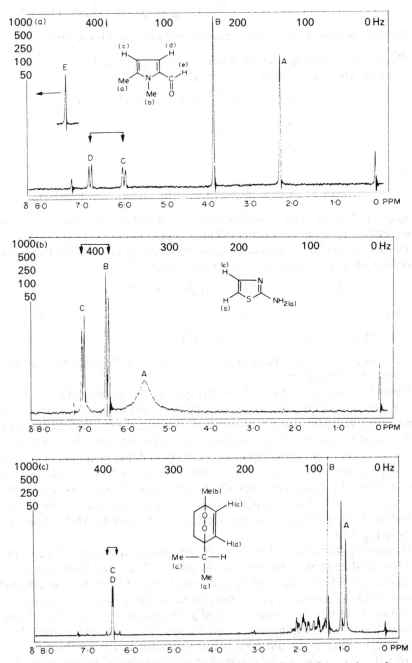

Fig. 3.18 The 60 MHz spectra of some compounds containing AB groupings of protons, in order of decreasing $v_0\delta/J$. The AB quartets are at low field in each case and are bracketed. (a) 1,5-Dimethylpyrrole-2-aldehyde. E is offset and is observed at 9.38 ppm. Protons (c) and (d) give the AB quartet. (b) 2-Aminothiazole. Note the broad line arising from hydrogen on ^{14}N. Protons (b) and (c) give the AB quartet. (c) Ascaridole. Protons (c) and (d) give the AB quartet. (Reproduced by permission of Varian International AG.)

interproton coupling constants, is not possible. This means that each proton has different and individual coupling constants to the three others. Thus in the fragment

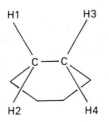

^3J(H1–H3) is not the same as ^3J(H2–H3) so that H1 and H2 (or H3 and H4) are not magnetically equivalent even if they have identical chemical shifts. This sort of situation is not simplified by increased operating frequency. We have thus reached a point in the history of NMR where, in many cases, the analysis of second-order spectra can be avoided by operating at a high enough frequency. In those cases where this will not work, computer programs are available that will permit the full analysis of virtually any system. We shall thus not consider further the systems that can be simplified, but shall look in detail at a typical system that is always second-order.

3.4.2   A four-spin $[AX]_2$ system

Typically, in inorganic chemistry, such systems occur in metal phosphine complexes such as *cis*-PdCl$_2${PF(OPh)$_2$}$_2$ where the nuclei A are the ^{31}P and the nuclei X are the ^{19}F. The remaining nuclei either have no influence on the ^{31}P or ^{19}F spectra, or their effect is removed by double irradiation. The notation $[AX]_2$ indicates that there two identical AX pairs of nuclei. The proton ring system of ascaridole is another example of this type. In the metal complexes such as the one chosen to illustrate this type of spectrum, there is strong spin–spin coupling

$$X–A \cdots A–X$$

between the directly bonded A and X nuclei and significant coupling between the distant A and X nuclei. There need be no coupling between the X nuclei, though in fact there will be a small, not necessarily observable, X–X coupling. There will be strong coupling between the A nuclei via the metal atom. It should be noted that the metal atom need not be present from the point of view of the NMR analysis of the system. An energy and transitions diagram of the system is shown in Fig. 3.19, set out so as to represent four AB-type diamonds for one nuclear species with the transitions in full lines. Each diamond shows transitions for the left-hand pair of spins (XX) for a fixed orientation of the right-hand (AA) pair, and these latter spin functions are marked in the centre of each

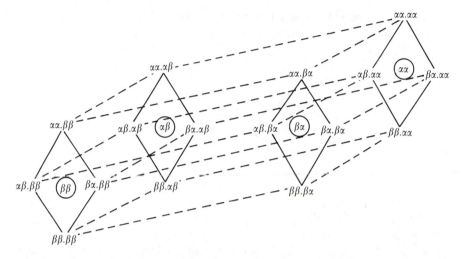

Fig. 3.19   Energy level and transition diagram for a four-spin [AX]₂ system. The spin functions α and β are shown in the order XX, AA. Each of the diamonds formed by full lines represents X transitions in the presence of fixed orientations of the AA nuclei, which are marked in the centre of each diamond.

diamond. The X spectrum is thus going to consist of four AB or AX quartets with possibly 16 resonances. However, there is some degeneracy of lines. Thus for the group of transitions ββ the A nuclei both have the same orientation and the X nuclei are isochronous. The effective chemical shift between them is then zero and we have the limiting singlet case of an AB spectrum. Similarly the group of transitions αα give a singlet X resonance. The frequencies of the two singlets are, of course, different because of the different A orientations and the difference is determined by the values of the long- and short-distance couplings to the A nuclei. It is thus $^1J(AX) + \,^3J(AX)$ for the present model system. Note that this spacing does not correspond to any actual coupling in the system but is composite (Fig. 3.20). For the remaining two groups of transitions, αβ and βα have the A nuclei in opposing orientations so that the X nuclei are not isochronous and the four lines of an AB spectrum are observed. Owing to the mixing of states described for the simple AB system, the effective coupling constant is not the rather small $^4J(XX)$ but is the much larger $^2J(AA)$. In the present case where we are dealing with a *cis* complex, $^2J(AA)$ is small relative to $^1J(AX)$ and we see two AB quartets disposed around each of the two main singlet peaks.

   Were we to investigate more complex spin systems, for example bis complexes of methyl phosphines, we would find the same general features. The methyl proton spectrum consists of a doublet plus several AB quartets, though the coupling constants are so much smaller that these are often not resolved from the main doublet peaks and the spectrum appears to be a simple doublet. In the

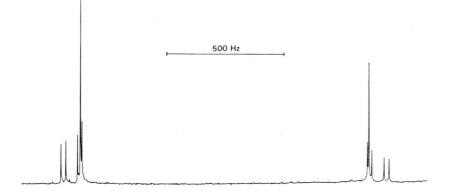

500 Hz

Fig. 3.20 Example of magnetic inequivalence in a chemically equivalent system. The ^{31}P spectrum of $cis\text{-}PdCl_2\{PF(OPh)_2\}_2$ in which the protons are decoupled is that of an $[AX]_2$ spin system, inequivalent because the P–F couplings are unequal. The spectrum is drastically different from the simple 1:2:1 triplet that would have been observed for an A_2X_2 spin system (From Mason (1987) *Multinuclear NMR*, Plenum, New York, with permission.)

case of *trans* phosphine ligands it is usual that $^2J(AA)$ is large relative to $^1J(AX)$, and in this case the resulting AB quartets are highly second-order with coupling apparently similar in value to that of the separation between the main doublet peaks. We thus obtain a closely spaced doublet at the centre of the spectrum and weak, unobservable flanking lines. Such spectra usually have the form of a triplet unless the resolution is particularly good, and are called deceptively simple spectra since the spacings do not correspond to any particular coupling and there are many almost degenerate lines. The example of Fig. 3.20 is particularly useful because it shows just how complex is the simplest four-spin system. Figure 3.21 shows the proton spectrum of a *trans* complex, *trans*-$PtBrMe(PMe_2R)_2$, already encountered in Fig. 3.13, where the effect of the ^{31}P atoms was removed by double irradiation. The full spectrum contains two principal triplets in intensity ratio 4:1 and flanked by similar triplet ^{195}Pt satellites. The smaller triplets near the TMS reference arise from the methyl ligand and the triplet splitting is due to equal coupling to the two phosphorus atoms in the phosphine ligands. This is, in fact, sufficient to identify the complex as being the *trans* form, since if the phosphine ligands were *cis* they would couple unequally to the methyl ligand, which would appear as a doublet of doublets. The methyl groups of the phosphine ligands cannot in any way couple equally to the two phosphorus atoms and their triplet is the deceptively simple triplet described above. Such triplets are taken to indicate *trans* structures for such complexes, whereas doublets arise from *cis* configurations.

An example of an organic molecule that has a four-spin $[AX]_2$ spectrum is

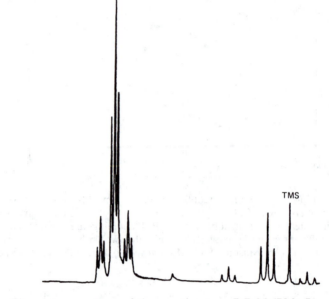

Fig. 3.21 The proton spectrum of the complex *trans*-PtBrMe(PMe$_2$R)$_2$ showing the two principal proton resonances in the intensity ratio 4:1. The flanking triplets are ^{195}Pt satellites and the small triplet splittings are due to coupling to the two ^{31}P atoms, the low-field group being deceptively simple in form. (Spectrum supplied by Professor B. L. Shaw.)

furan, whose proton spectrum is shown in Fig. 3.22 together with a formula provided with the various proton spin–spin coupling constants. The resonances of the α and β protons appear at first glance to be triplets, but close inspection at high resolution shows the central line to be composite. A full analysis of the system, shown by the stick diagrams, gives a doublet spectrum plus a pair of AB sub-spectra with most intensity in the central resonances. Interestingly, the ^{13}C spin satellites of this molecule are first-order in appearance. The presence of a ^{13}C atom in a molecule splits the resonance of the attached proton into a widely separated doublet, so that it is no longer isochronous with its twin. This introduces an effective chemical shift, which permits these hydrogen atoms to couple with each of the other three and give simplified spectra that can aid interpretation of the main spectrum.

Questions

3.1. Figure 3.6 shows the ^{1}H spectrum of an ethyl group. Measure the chemical shift of the two multiplets and their coupling constant.

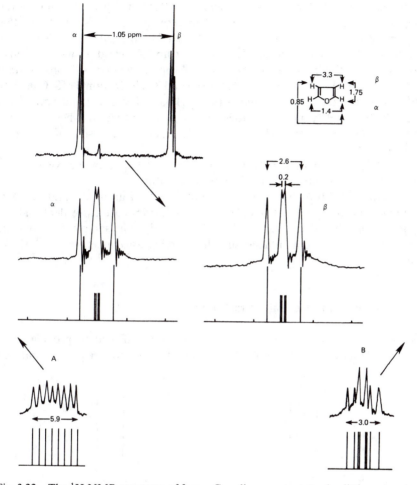

Fig. 3.22 The ^{1}H NMR spectrum of furan. Coupling constants and splittings are given in hertz. The A satellite is 100.7 Hz to low field of the α multiplet and the B satellite 87.6 Hz to high field of the β multiplet. $^1J(H-^{13}C) = 201.4$ Hz and $^1J(H-^{13}C) = 175.3$ Hz, respectively. (From Reddy and Goldstein (1962) *J. Am. Chem. Soc.*, **84**, 583; copyright (1962) American Chemical Society, reprinted with permission.)

3.2. Verify that the ratio between the coupling constants to ^{1}H of the nuclei ^{10}B and ^{11}B in the BH_4^- are in the ratio of the magnetogyric ratios of the two boron isotopes (Fig. 3.8).

3.3. Figure 3.9 shows the proton spectrum of a CD_2H group in isotopically substituted acetone. $^2J(H-D) = 2.4$ Hz. Calculate the coupling constant $^2J(H-H)$ between the protons in the CH_3 groups of normal acetone, ignoring isotope effects on J.

3.4. Doublet satellite signals are observed in the ^{1}H spectrum of acetone, close to the main singlet signal, with a spacing of 5.9 Hz, which is the value of the coupling $^2J(^1H-^{13}C)$. What will be the pattern in the ^{13}C spectrum of this carbonyl group, i.e. number of lines and their relative intensities.

3.5. Figure 3.18(c) shows the 60 MHz ^1H spectrum of ascaridole, which features a tightly coupled AB quartet at about $\delta = 6.45$ ppm. The line positions, starting at the lowest-field one, are 395.5, 386.9, 385.5 and 376.9 Hz from TMS. Calculate $^3J(Hc-Hd)$ and the chemical shift between Hc and Hd in Hz and ppm. Then calculate the relative line intensities and positions expected when the same sample is observed at 500 MHz. What is the exact chemical shift of the centre of the quartet.

3.6. Explain why, in Figs. 3.18(a) and (b), the two halves of the AB quartets have different intensities and so different linewidths.

3.7. Figure 3.23 shows the ^{19}F resonance of the CF_2H group of 1, 1, 1, 2, 2, 3, 3-heptafluoropropane, $CF_3CF_2CF_2H$. The two fluorines are equivalent and are coupled to all the other magnetically active nuclei in the sample. Pick out the various multiplet patterns and measure the coupling constants $^2J(F-H)$, $^3J(F-F)$ and $^4J(F-F)$.

Exercises in spectral interpretation

The proton spectra of simple molecules are often sufficient to provide a full structural description of the molecule. The task is made even easier if an infrared

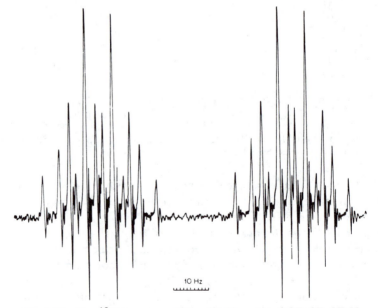

10 Hz

Fig. 3.23   The ^{19}F resonance of the CF_2H group of $CF_3CF_2CF_2H$.

(IR) spectrum is also to hand, since the two often give complementary information. Since we are, however, concentrating on NMR here, we will ignore the IR and attempt to obtain the maximum from the NMR traces.

An NMR spectrum contains several pieces of usable information. First of all is the chemical shift. Thus the position of a resonance indicates the type of group in which the protons reside, sometimes with remarkable clarity, though often there will be ambiguities. The chemical shift ranges within which several types of proton are found are given in the accompanying chemical shift chart (Fig. 3.24). It is only necessary to add that hydrogen-bonded protons are found to low field, often below 10 ppm, and that metal alkyls are found to high field around and above TMS. Hydrogen bonded to metals occurs in a wide range also above TMS and varies from about -3 ppm, e.g. $HRe(PPh_3)_3(CO)_2$, to -50 ppm, e.g. $HIrCl_2(phosphine)_2$, though there are one or two exceptions with shifts to low field of TMS. The second piece of information is the integral trace of the spectrum, which gives the area under each resonance and so the relative numbers of protons contributing to each resonance. This, coupled with the empirical formula, will enable the hydrogen in the molecule to be split up into chemically different subgroups. Spin–spin coupling patterns also give this type of quantitative information, with the difference that an ethyl quartet–triplet pattern is diagnostic but does not tell us how many ethyl groups are present. The existence of coupling also tells us that the coupled groups are proximate.

The interpretation of spectra is thus a deductive process in which one attempts to account for all the spectral features in terms of a single molecular structure. We will work through some examples and then provide the reader with some exercises to try. We will start with a few simple singlet spectra where the only information is the chemical shift and the formula:

1. The compound has a singlet at 7.27 ppm and formula C_6H_6. Evidently we have sample of benzene.
2. The compound produces a singlet at 2.09 ppm and has a formula C_3H_6O. The chemical shift is typical of methyl or methylene in unsaturated molecules or of CH_3CO. Again, it is a short step to acetone $(CH_3)_2CO$.
3. The compound gives two singlets of equal intensities at 2.01 and 3.67 ppm and has a formula $C_3H_6O_2$. Clearly, the two singlets arise from three protons each, which form two equivalent sets and which are not coupled. There are thus two CH_3- and by difference we have CO_2. The chemical shifts are typical of CH_3O and CH_3CO. Thus we have the ester CH_3COOCH_3, with four bonds between the hydrogen atoms in the two groups and so no detectable coupling.

The student should also compare the assigned resonances in Fig. 3.18 with the chemical shift chart in Fig. 3.24, since these cover a wide range. Note that NH_2 resonances are very variable in position.

Now we will consider an example where there is spin–spin coupling. The spectrum shown in Fig. 3.25 is that of the substance $C_3H_5O_2Cl$. It consists of nine resonances, of which we can discount two as arising from the reference

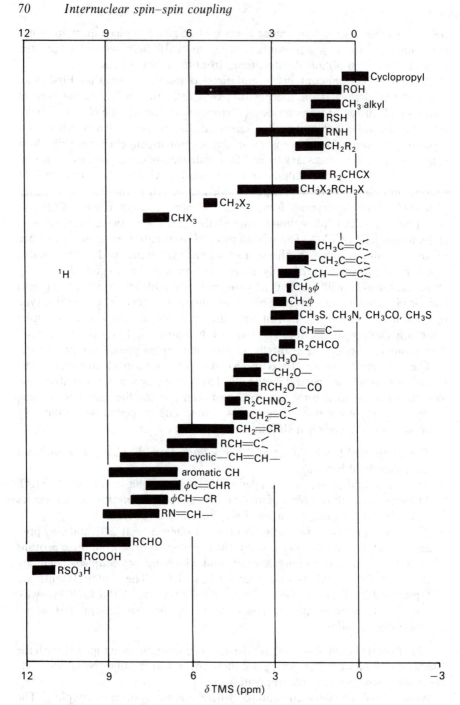

Fig. 3.24 Chart of approximate chemical shift ranges of different types of protons in organic compounds. X = halogen, R = organic substituent, φ = phenyl. (From Akitt (1987) in *Multinuclear NMR*, ed. Mason, Plenum, New York, with permission.)

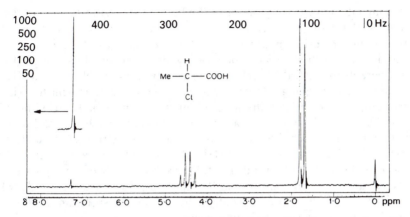

Fig. 3.25 The 60 MHz ^{1}H spectrum of $C_3H_5O_2Cl$. (Reproduced by permission of Varian International AG.)

TMS (0 ppm) and the remnant $CHCl_3$ in the deuterated $CDCl_3$ solvent (7.3 ppm). Two of the resonances are split into regular multiplets with identical splitting of 7.3 Hz. This informs us that there is spin–spin coupling between two of the groups of protons and the value of the coupling constant is typical of a vicinal three-bond coupling pathway, though not exclusively so. The doublet–quartet pattern must arise from a $H-H_3$ interaction. The integrals are in the ratio 3(doublet):1(quartet):1(singlet), and since there are five protons in the molecule the hydrogen atoms must also be split up in this ratio. The chemical shift of the doublet is typical of CH_3 alkyl; that of the quartet is rather low-field for CHX but we need also to keep in mind the influence of the oxygen-containing part of the molecule, which is also to low field. Finally, the singlet appeared too far to low field to fit on the chart and had to be offset. This is typical for acidic protons or aldehydes. We can now attempt to work out the structure, and it helps to do this if we settle upon one structural unit at a time and subtract this from the formula. Thus we have $C_3H_5O_2Cl$. The doublet is evidently a CH_3 unit. This leaves $C_2H_2O_2Cl$. It is coupled to an alkyl proton, i.e. a CH, to give CH_3CH, leaving CHO_2Cl. Two valencies on the CH carbon need to be filled. Cl must take one, leaving CHO_2, and this carboxyl group must then take the other. The molecule is

Now apply the same approach to exercises 1 to 12. Bear in mind the following: (a) The fluorine resonances of the compounds containing fluorine are not visible but the proton resonances may be coupled to the ^{19}F and so split into multiplets; the same comments apply also to the phosphorus compound. (b) Also, broadened NH proton resonances may still cause splitting in vicinal protons, the broadening hiding the corresponding multiplicity of the NH resonance. (c) The outer lines of a multiplet may not be visible; use Pascal's triangle to check the relative line intensities in at least one of the exercises. (d) Alcoholic protons are variable in position because of differing hydrogen bonding effects. The spectra were all obtained at 60 MHz, though a few bear inserts taken at 100 MHz, which allow spin coupling and chemical shift effects to be distinguished. The numbers on the spectra give the relative numbers of protons contributing to each resonance and have been obtained from integral traces. Exercises 1 and 7 show expanded regions of the spectra so as to allow fine structure to be distinguished. These were obtained at 60 MHz also. The answers are given at the end of the section, though several structures are given, only one being correct. *All* the features of a spectrum should be explicable from the structure. It will be instructive also to consider what differences in the spectrum would be obtained from the incorrect structures given.

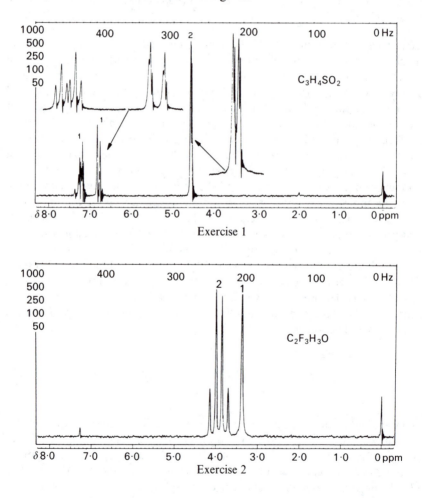

Exercise 1

Exercise 2

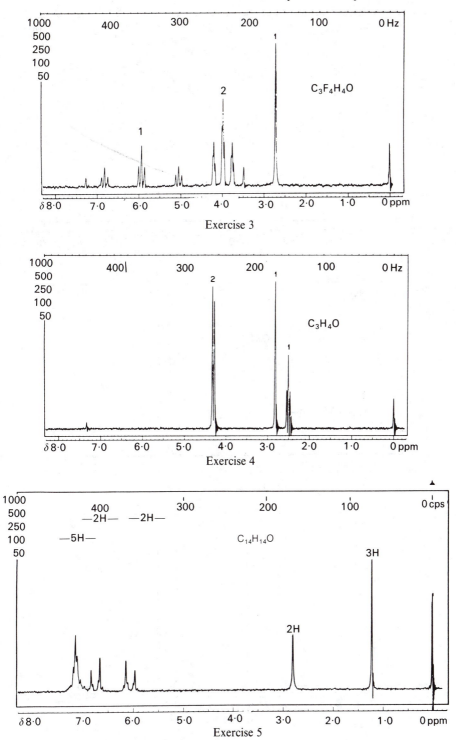

Exercise 3

Exercise 4

Exercise 5

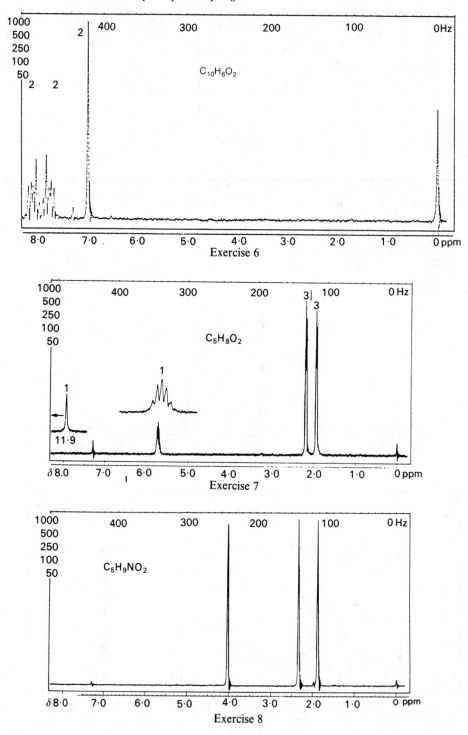

Exercise 6

$C_{10}H_6O_2$

Exercise 7

$C_5H_8O_2$

11·9

Exercise 8

$C_5H_9NO_2$

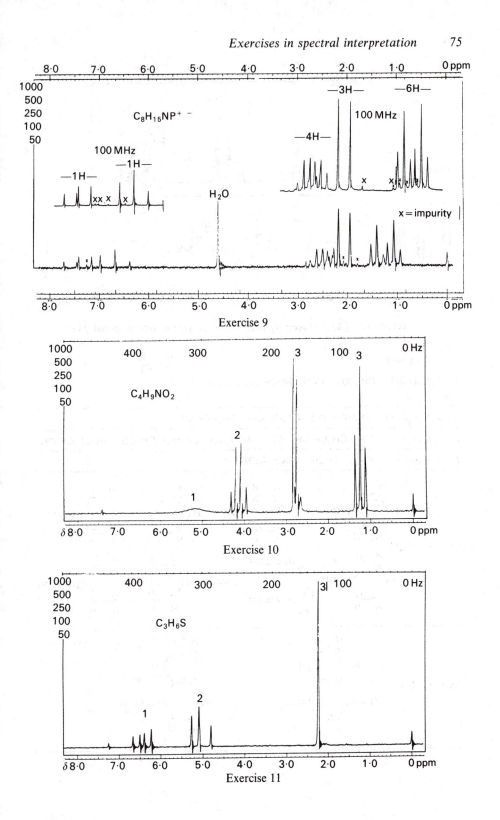

Exercise 9

Exercise 10

Exercise 11

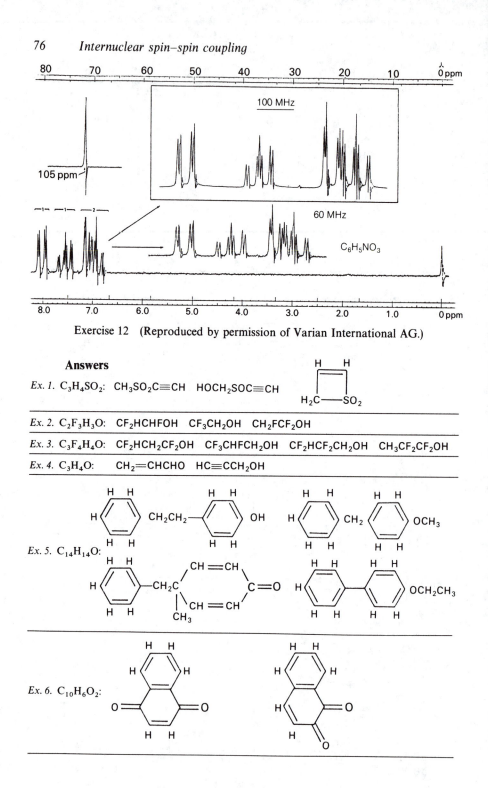

Exercise 12   (Reproduced by permission of Varian International AG.)

Answers

Ex. 1. $C_3H_4SO_2$: $CH_3SO_2C\equiv CH$ $HOCH_2SOC\equiv CH$

$$\underset{H_2C\text{———}SO_2}{\overset{H\quad\quad H}{\square}}$$

Ex. 2. $C_2F_3H_3O$: $CF_2HCHFOH$ CF_3CH_2OH CH_2FCF_2OH

Ex. 3. $C_3F_4H_4O$: $CF_2HCH_2CF_2OH$ CF_3CHFCH_2OH $CF_2HCF_2CH_2OH$ $CH_3CF_2CF_2OH$

Ex. 4. C_3H_4O: $CH_2{=}CHCHO$ $HC\equiv CCH_2OH$

Ex. 5. $C_{14}H_{14}O$:

Ex. 6. $C_{10}H_6O_2$:

Ex. 7. $C_5H_8O_2$: $(CH_3)_2C$=$CHCOOH$ CH_3CH=$CCOOH$ CH_3CCH=$CHCH_2OH$

$$\underset{\displaystyle CH_3}{|} \qquad \underset{\displaystyle O}{\|}$$

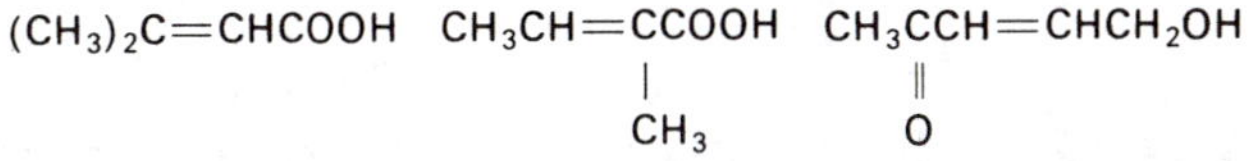

Ex. 8. $C_5H_9NO_2$: $ONC(CH_3)_2COCH_3$ CH_3ON=$C(CH_3)COCH_3$ $(CH_3)_2C$=$C(CH_3)NO_2$

Ex. 9. $C_8H_{15}NP^+I^-$: $H_3\overset{+}{P}C(CH_2CH_3)_2CH$=$CHCN$ $\underset{CH_3CH_2}{\overset{CH_3CH_2}{\underset{\diagup}{\diagdown}}}\overset{+}{P}\underset{CH=CHCN}{\overset{CH_3}{}}$ $\underset{CH_3CH_2}{\overset{CH_3CH_2}{\underset{\diagup}{\diagdown}}}\overset{+}{P}\underset{CH=CH_2}{\overset{CH_2CN}{}}$

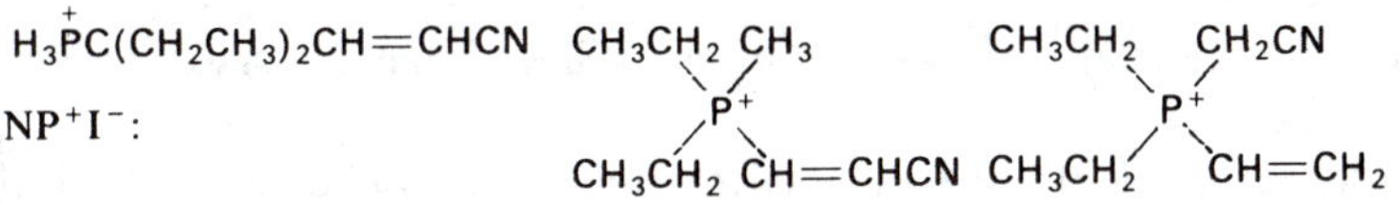

(Compare the values of $J(PH)$ to the methylene protons, the ethylene protons and the triplet of protons.)

Ex. 10. $C_4H_9NO_2$: $CH_3NHCOOCH_2CH_3$ $H_2NCH_2CH_2CH_2COOH$ $CH_3CH_2CHCOOH$

$$\underset{\displaystyle NH_2}{CH_3CHCOOCH_3 \atop |} \qquad \underset{\displaystyle CH_3}{CH_3COCH_2\,NOH \atop |} \qquad \underset{\displaystyle NH_2}{|}$$

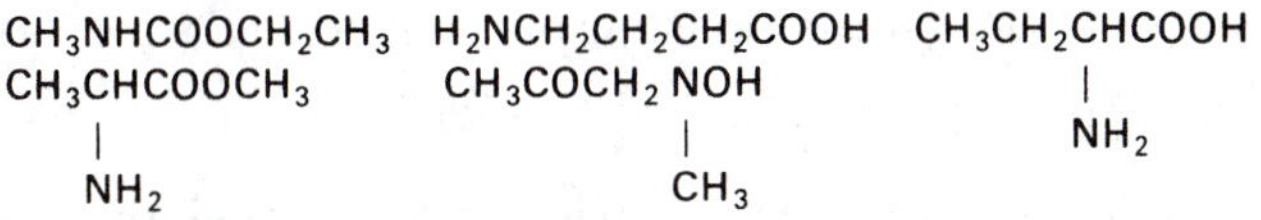

Ex. 11. C_3H_6S: CH_3CH=$CHSH$ CH_2=$CHSCH_3$ CH_2=CCH_3SH

Ex. 12. $C_6H_5NO_3$:

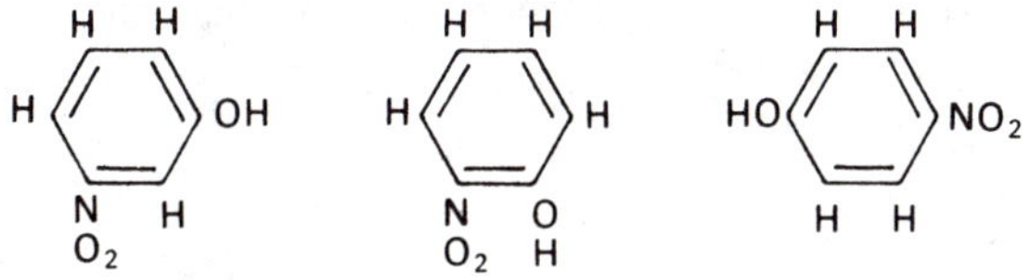

4 Nuclear magnetic relaxation

4.1 Relaxation processes in assemblies of nuclear spins

If we perturb a physical system from its equilibrium condition and then remove the perturbing influence, the system will return to its original equilibrium condition. It does not return instantaneously, however, but takes a finite time to readjust to the changed conditions. The system is said to relax. Relaxation to equilibrium usually occurs exponentially, following a law of the form

$$(n - n_e)_t = (n - n_e)_0 \exp(-t/T)$$

where $(n - n_e)_t$ is the displacement from the equilibrium value n_e at time t and $(n - n_e)_0$ that at time zero. The relaxation can be characterized by a characteristic time T. If T is small, relaxation is fast; whereas if T is long, relaxation is slow. In recent years it has become common to speak also of rates of relaxation, which are given the symbol R. Evidently

$$R = 1/T$$

R has the advantages that its value increases as the relaxation becomes faster and that, if a system is subject to several parallel relaxation processes, then the overall rate is the sum of the rates of all the processes, i.e.

$$R = R_a + R_b + R_c + \cdots$$

The relaxation behaviour of assemblies of nuclear spins shows up directly in their NMR spectra and is related to the molecular dynamics of the system. For these reasons it is of considerable importance to NMR spectroscopists, on the one hand allowing them to optimize experimental conditions, even to eliminate some undesirable spectral feature, or on the other hand allowing close study of the physical and chemical properties of the motion of a system.

We have already seen that in an assembly of spin-1/2 nuclei immersed in a strong magnetic field the spins are polarized into two populations with opposite senses and with a small excess number in the lower energy state. The nuclei precess around the magnetic field direction with a net magnetization M_z and no detectable transverse magnetization in the xy plane. It turns out that this system can be perturbed in two ways, and that we have to expect that there may be two relaxation processes with different relaxation times, which we will call T_1 and T_2, or rates of relaxation R_1 and R_2. We have already seen in Chapter 1 (Fig. 1.6) that a B_1 radiofrequency pulse can swing the total nuclear

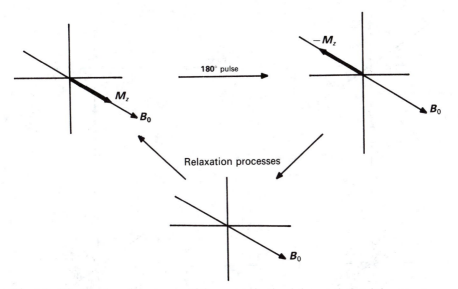

Fig. 4.1   The T_1 relaxation process. If the magnetization is inverted, then it has to return to its equilibrium state and does so by decaying to zero and then increasing again in the normal B_0 direction. This process involves an exchange of energy between the spins and their environment.

magnetization away from its equilibrium position in the z direction, and this is essentially a perturbation of the system. If we apply a rather long pulse we can swing the magnetization back into the z direction but pointing in the opposite direction. Such a pulse is called a 180° pulse, the reason for this name being evident in Fig. 4.1. The magnetization has been inverted and, immediately following the end of the pulse, relaxation processes start to return the magnetization to its normal state. Thus the magnetization decays to zero and then builds up to attain its normal value and direction. The characteristic time for this process is T_1. The process is also called longitudinal relaxation since it takes place in the direction of B_0, and in solid materials it is also called spin–lattice relaxation. In all cases it must be emphasized that the inverted magnetization has higher energy than the normal magnetization and that the return to equilibrium involves an exchange of magnetic energy with the surroundings – the lattice in the case of a solid. A simple way to demonstrate the T_1 process is to take a sample of ethanal (acetaldehyde; used for checking spectrometer resolution) and place it in the sample space of a spectrometer that is scanning a spectrum every 2 or 3 s. The T_1 of this sample is quite long and the signal strength can be observed to increase as the nuclei become polarized in the magnetic field.

If instead we use a 90° pulse to perturb the spin system, we move the magnetization into the xy plane as in Fig. 4.2. Now the magnetization in the z (B_0) direction is zero, and this returns to its normal value by the mechanism just

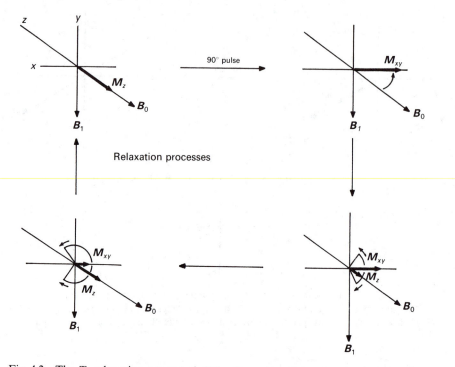

Fig. 4.2   The T_2 relaxation process. A 90° pulse swings the magnetization into the xy plane around the B_1 vector. This is shown as stationary in the figure as if the observer were rotating in the same direction at the Larmor frequency, thus giving a static picture. There is a spread of nuclear frequencies, which causes the spins to fan out and reduce the resultant M_{xy}. M_z increases at the same time due to the T_1 process.

discussed. However, we have also created transverse magnetization in the xy plane, which rotates at the nuclear Larmor frequency. This has to decay to zero at equilibrium and does so because the frequency of each spin differs slightly from that of its companions, each varying randomly around the mean precession frequency. This means that, on average, some spins are slower and some are faster than the mean, so that the xy magnetization starts to fan out, to lose coherence and the resultant to become less in magnitude. Eventually the spins take all directions in the xy plane and M_{xy} is zero. The characteristic time of this process is T_2. This is also called transverse relaxation, and in solid materials it is known as spin–spin relaxation. Since the process is related to the spread of frequencies of a nuclear resonance, it is evident in the spectral linewidth. T_1 and T_2 may be equal or they may differ by orders of magnitude. The T_2 process involves no energy change. Evidently M_z increases as M_{xy} decreases, so that T_1 and T_2 can be equal, but T_2 cannot be longer than T_1. On the other hand, T_2 can be less than T_1, and in this case M_{xy} decays more rapidly than M_z is re-established and the signal (derived from M_{xy}) disappears well before equilibrium of the spins is attained.

The nuclei of atoms are extremely well isolated from their surroundings and, because the energy of NMR transitions is small, the chance of a spin transition occurring spontaneously is negligibly small. The fact that relaxation times can be quite short indicates that transitions are stimulated, and we must thus consider the various ways that this can happen.

4.2   Dipole–dipole relaxation

We have already remarked in section 2.1 that the magnetic field at a given nucleus due to the magnetic moment of a near-neighbour nucleus is very high but is averaged to zero by the random rotational diffusion of the molecule in which the nuclei reside. The magnitude of this field is such that the instantaneous chemical shift displacement of one ^1H nucleus due to the other in a methylene group can be as high as $150\,000\,Hz$. As the group rotates, this field varies by such an amount on each side of zero. Thus the nuclei have instantaneously different precession frequencies since all possible orientations of the molecules will exist at any one instant. Randomization of the frequencies means not that all will have the same frequency in the long term but rather that once out of step a nucleus is just a probable to move further away from its companion's frequency as to reconverge to it. This dipole–dipole fluctuating field then is the cause of the loss of coherence between spins and so the source of the T_2 relaxation process.

The chaotic random motion of a solute in a solvent is called Brownian motion. This has a timescale that depends upon a number of factors such as mass of solute, solution viscosity and temperature. Because the motion is random, this timescale is characterized by a somewhat loosely defined term, the rotational correlation time τ_c. This is the time taken on average for a solute molecule to rotate by one radian, or, more precisely, the time interval after which the molecular motion contains no vestige of its earlier angular momentum, that is, has lost all memory of its previous behaviour. This time τ_c is typically $10^{-11}\,s$ in liquids of low viscosity, which converts on the frequency scale to $10^5\,MHz$. This is around the maximum rate of motion in the system, and all slower rates of motion can exist. The frequency spectrum of such random motion and associated magnetic fields is simple and is essentially white noise at all frequencies less than $1/\tau_c$. Since the maximum NMR frequency that is likely to be encountered is $600\,MHz$, it is clear that there is a component at all possible NMR frequencies. The relaxation field thus provides a B_1 component that varies in intensity and direction and causes random precession of the nuclei also: hence the dephasing of spins and also the possibility of energy transfer needed for the T_1 mechanism to operate.

The field intensity at any frequency, $K(v)$, is given by

$$K(v) \propto \frac{2\tau_c}{1 + 4\pi^2 v^2 \tau_c^2} \tag{4.1}$$

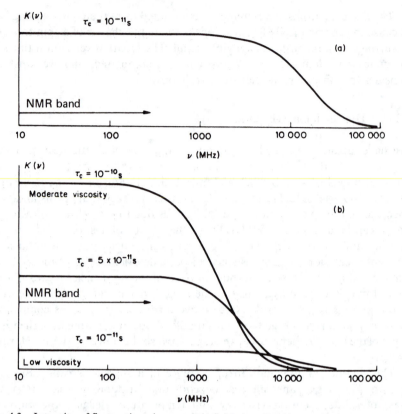

Fig. 4.3   Intensity of fluctuations in magnetic fields in a liquid sample due to Brownian motion, as a function of frequency; (a) typical liquid; (b) liquids with different correlation times.

This function is plotted in Fig. 4.3, where it will be seen that over the NMR frequency band the intensity of the relaxation field is constant. However, if τ_c is rather longer, this may not be true. The Debye theory of electric dispersion shows that, for a spherical molecule rotating in a liquid, the correlation time is given by

$$\tau_c = \frac{4\pi a^3}{3k}\frac{\eta}{T}$$

where η is the viscosity, of the liquid, T is the temperature and a is the radius of the molecule. Thus if we vary the viscosity of a sample, or its temperature, or the mass of the solute molecule, we will change τ_c. The effect of varying τ_c is shown also in Fig. 4.3, where it will be evident that increasing τ_c increases the intensity of the relaxation field, though for the highest NMR frequencies and the longest correlation times shown, we start to leave the flat portion of the spectrum and the relaxation field starts to decrease again. Provided we limit our range to the flat portion of these curves, then the relaxation field and so the relaxation rate increases with τ_c. Long relaxation times thus occur for low viscosity, high temperature and small molecular mass.

It will be seen from equation (4.1) that the relaxation field intensity has its flat frequency response when the quantity $4\pi^2\nu^2\tau_c^2$ is very much less than unity, i.e.

$$4\pi^2\nu^2 \ll 1/\tau_c^2$$

This is known as the region of extreme narrowing, where the correlation time is much shorter than one Larmor period of the nuclei.

We can now make some qualitative predictions about how the relaxation times will vary with τ_c. In the extreme narrowing limit, T_1 and T_2 are determined by the same relaxation field and so are equal. Increasing τ_c reduces the relaxation times until we reach the point at which the frequency spectrum of the field is no longer flat. Its intensity begins to decrease at the higher NMR frequencies, and so the chance for energy exchange is decreased and T_1 increases with further increase in τ_c. There is thus a minimum in T_1. The behaviour of T_2 is quite different since the correlation time is now similar in length to a Larmor period, and superimposed on the random field we see for short periods the rotating vector of the neighbouring nuclear magnetic moment. This is at the nuclear frequency and provides a second means for loss of coherence in the xy plane. T_2 continues to decrease as τ_c increases. In the limit of infinite τ_c we have the solid state. Here there is no dipole–dipole relaxation field and T_1 is very long and is determined by the presence of ferromagnetic impurities in the lattice. Hence the name spin–lattice relaxation. T_2 is determined by the now vary strong interaction between spins via the rotating field generated by the Larmor precession of the spins. Rapid exchange of spin states is stimulated, the lifetime of the individual spin state is short and the uncertainty principle dictates short T_2. Hence also the name spin–spin relaxation.

A full analysis of the spectral density of the relaxation field and the way this influences the spins gives the following equations for T_1 and T_2, expressed as the rates. We will also introduce the notation R_{1DD} and R_{2DD} or T_{1DD} and T_{2DD} to indicate that the mechanism discussed is dipole–dipole interaction.

For two *identical* spin-1/2 nuclei situated in the same molecule, the intra-molecular dipole–dipole relaxation rates are

$$R_{1DD} = 2a\frac{\gamma^4}{r^6}\left(\frac{\tau_c}{1+\omega^2\tau_c^2} + \frac{4\tau_c}{1+4\omega^2\tau_c^2}\right)$$

$$R_{2DD} = a\frac{\gamma^4}{r^6}\left(3\tau_c + \frac{5\tau_c}{1+\omega^2\tau_c^2} + \frac{2\tau_c}{1+4\omega^2\tau_c^2}\right)$$

where $a = 3\mu_0^2\hbar^2/320\pi^2$, μ_0 is the permeability of a vacuum, r is the distance between the nuclei and γ is their magnetogyric ratio. The way these rates of relaxation vary with τ_c is shown on Fig. 4.4 for two protons at two different

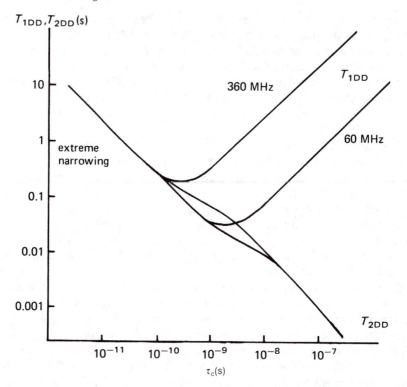

Fig. 4.4   Variation of T_{1DD} and T_{2DD} with τ_c for two different spectrometer frequencies. The figures given apply to two protons separated by 160 pm. (After Martin *et al.*, *Practical NMR Spectroscopy*; copyright (1980) John Wiley and Sons Inc., New York, reprinted with permission.)

spectrometer frequencies ω. The main feature to note is the T_1 minimum, which marks the limit of the extreme narrowing region and the way this moves with spectrometer frequency. The higher this frequency, the shorter becomes the maximum permitted τ_c, the result being that, for large complex molecules where the highest frequencies may be needed to give the necessary degree of resolution, the increased correlation times of such molecules may result in reduced T_2 and so increased linewidths. The expression for R_{1DD} and R_{2DD} is much simpler in the extreme narrowing region:

$$R_{1DD} = R_{2DD} = 10a\frac{\gamma^4}{r^6}\tau_c \qquad (4.2)$$

Two further points should be emphasized. In the first place the rate of dipole–dipole relaxation depends upon the fourth power of the magnetogyric ratio, and nuclei with large magnetic moments will be most strongly subject to such relaxation, for example, ^1H or ^{19}F. The mechanism will be of lesser importance for nuclei with smaller magnetogyric ratios. Secondly, the efficiency of relaxation depends strongly upon the inverse distance between the spins. The effect of more distant spins will be almost negligible.

For two *different* spin-1/2 nuclei in the same molecule with magnetogyric ratios γ_I and γ_S respectively, the intramolecular dipole–dipole relaxation rates of one (species I) due to interaction with other (species S) are

$$R_{1DD} = \frac{a}{3} \frac{\gamma_I^2 \gamma_S^2}{r^6} \left(\frac{2\tau_c}{1 + (\omega_I - \omega_S)^2 \tau_c^2} + \frac{6\tau_c}{1 + \omega_I^2 \tau_c^2} + \frac{12\tau_c}{1 + (\omega_I + \omega_S)^2 \tau_c^2} \right)$$

$$R_{2DD} = \frac{R_{1DD}}{2} + \frac{a}{3} \frac{\gamma_I^2 \gamma_S^2}{r^6} \left(4\tau_c + \frac{6\tau_c}{1 + \omega_S^2 \tau_c^2} \right)$$

These equations, though more complex, give plots of form very similar to those depicted in Fig. 4.4. In the extreme narrowing limit we find

$$R_{1DD} = R_{2DD} = \frac{20a}{3} \frac{\gamma_I^2 \gamma_S^2}{r^6} \tau_c \tag{4.3}$$

The most common example of relaxation of one spin by a different species occurs in ^{13}C spectroscopy, where the carbon nucleus is relaxed by the moments of directly bonded hydrogen atoms. In addition, because the relaxation field is stronger if more hydrogen atoms are attached to the ^{13}C, the rates of relaxation are proportional to the number of directly bonded hydrogens. Thus for relaxation of ^{13}C in a CH_3 group, the above expression has to be multiplied by 3. Provided that relaxation occurs solely by the dipole–dipole mechanism, and the C–H bond length is known, then the correlation time may be calculated from the measured relaxation time. Relaxation of tertiary or carbonyl ^{13}C is much slower than for CH_n groups since the C–H distance is much greater, and in this case other relaxation mechanisms may be important. The relaxation of protons on ^{13}C is dominated by interaction with other protons in the molecule.

4.3   Quadrupolar relaxation

Relaxation times for protons in organic molecules, as will be seen from Fig. 4.4, are of the order of seconds. For spin-1/2 nuclei with small magnetogyric ratios, the times are in general much longer, particularly in the absence of neighbouring protons. If the nucleus in question, however, has spin $> 1/2$, it has a quadrupole moment, and this introduces a second and very efficient relaxation mechanism, which results in a generally much reduced relaxation time, a factor of as high as 10^8 over the expected dipole–dipole relaxation time being possible.

The quadrupole moment of a nucleus arises because the distribution of charge is not spherical, as is the case for spin-1/2 nuclei, but is ellipsoidal, i.e. the charge distribution within the nucleus is either slightly flattened (oblate – like the Earth at its poles) or slightly elongated (prolate – like a rugby ball). Cross-sections of the charge of two such nuclei are shown in Fig. 4.5 with the departure from

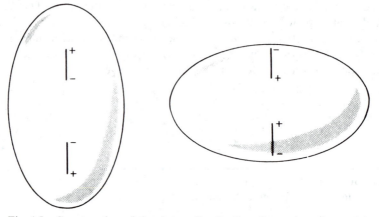

Fig. 4.5   Cross-section of the charge distributions in quadrupolar nuclei.

spheroidal form much exaggerated. The diposition of the electric quadrupole is also suggested. If the electric fields due to external charges vary across the nucleus, then the torque on each dipole component of the quadrupole is different, and a net torque is exerted on the nucleus by the electric field as well as by the magnetic fields present. Electric field gradients exist at atomic nuclei due to asymmetries in the spatial arrangement of the bonding electrons. The Brownian motion of the molecule causes the direction of the resulting electric quadrupole torque to vary randomly around the nucleus in exactly the same way as does the torque due to the magnetic relaxation field. Rotating electric torque components thus exist at the nuclear resonant frequency, which can also cause interchange of energy between the nucleus and the rest of the system (T_1 mechanism) and randomization of nuclear phase (T_2 mechanism).

The electric field gradient at a point in a molecule arises from the effect of the surrounding electric charges. The electric field can be described by three components in a Cartesian system of axes, but the field magnitude will change as one moves along an axis and in general all three of its components will change. Further, the changes may be different depending on which axis one chooses. The electric field gradient (EFG) thus requires nine numbers in order to describe it fully, and is a tensor quantity. Fortunately, by choosing an appropriate system of axes relative to the molecular geometry, this number can be reduced to three, V_{xx}, V_{yy} and V_{zz}, the diagonal components of the tensor, which have the further property that their sum is zero. This permits a further simplification by introducing the asymmetry factor, η, such that

$$\eta = (V_{yy} - V_{xx})/V_{zz}$$

V_{zz} is chosen to be the largest component of the EFG tensor, and this can then be described by two quantities, V_{zz} and η. If the system is axially symmetric so that $V_{yy} = V_{xx}$ then $\eta = 0$. The V_{xx}, etc., are calculated as the sum of contributions from all charges i:

$$V_{xx} = \sum_i q_i r_i^{-5}(3x_i^2 - r_i^2) \tag{4.4}$$

where r_i is the distance of charge q_i from the nucleus and x_i is the coordinate of the charge in the axis system chosen. Similar expressions give V_{yy} and V_{zz}.

The equation describing the quadrupole relaxation times (T_{1Q}, T_{2Q}) of a quadrupolar nucleus with spin I situated in a molecule with an isotropic correlation time τ_c and in the extreme narrowing region (expressed as the rates) is

$$R_{1Q} = R_{2Q} = \frac{3\pi^2}{10} \frac{(2I+3)}{I^2(2I-1)} \left(\frac{eQ}{h}\right)^2 V_{zz}^2(1 + \tfrac{1}{3}\eta^2)\tau_c \tag{4.5}$$

where Q is the quadrupole moment of the nucleus I, e is the electronic charge and h is Planck's constant. Outside the region of extreme narrowing, the behaviour is reminiscent of that of dipole–dipole relaxation but is complicated by the fact that several nuclear transitions are possible. Thus for a nucleus with spin 3/2 there are three transitions, $3/2 \leftrightarrow 1/2$, $1/2 \leftrightarrow 1/2$ and $-1/2 \leftrightarrow -3/2$, and the relaxation rate for nuclei undergoing the first and last transitions differs from that of those undergoing the $1/2 \leftrightarrow -1/2$ transition. This produces what is called non-exponential relaxation.

Quadrupolar relaxation depends strongly upon both nuclear properties (Q, I) and molecular properties (V_{zz}, η, τ_c). Its effectiveness increases rapidly as Q is increased, though this is to some extent offset by the fact that large Q tends to be associated with large I and the function of I in the expression for T_{1Q} decreases rapidly as I increases. As a result, almost all quadrupolar nuclei are capable of being detected, only a few having such short relaxation times that they are unobservable. Of the molecular properties, the correlation time has a particularly strong influence in that changes in τ_c brought about by changes in temperature alter the relaxation times considerably. Increasing temperature reduces τ_c and so increases the relaxation time and reduces the resonance linewidth. Since the resonances of quadrupolar nuclei are usually quite broad, these changes are evident. The same effect occurs with dipole–dipole relaxation, but here the relaxation times are long and the linewidth is determined more by the spectrometer resolution than by the relaxation time, and the changes are not so immediately obvious.

For a given nuclear species, the quadrupolar relaxation time is determined mainly by the EFG and this can vary from almost zero to very large values, so that relaxation times can vary by several orders of magnitude depending upon the situation of the nucleus. In principle, the EFG can be calculated at any point in a molecule so that one could obtain τ_c, or if τ_c were known (e.g. from the ^{13}C relaxation time of a CH moiety in the molecule) the EFG could be determined and compared with that calculated so as to verify the accuracy of the calculation. Unfortunately, it proves in practice difficult to calculate the EFG with sufficient accuracy. Equation (4.4) shows that the EFG is proportional to the inverse cube of the distance of the charge from the nucleus. Thus charge close to the nucleus has the predominating effect, and it is here that the calculations are least accurate. In fact, there is some confusion in the literature on this point. In the solid state, the EFG arises from quite distant charges as well as those close by, in the same way that a Madelung constant is calculated. In a liquid, the movement of the molecules reduces the distant effects to zero and the EFG arises quite locally around the nucleus. For instance, in the anion $AlCl_3(NCS)^-$ both the quadrupolar nuclei ^{27}Al and ^{14}N have long relaxation times and show spin–spin coupling. There are two points of low EFG in the molecule, and this can only arise if the EFG arises in the region quite close to the nucleus. The question is, of couse, how close has the charge to be, to be effective?

The cases of greatest interest are those in which the relaxation times are relatively long, since this gives the best resolution of resonances and the possibility of seeing coupling effects. This means that we should understand the requirements for obtaining a low EFG at a nucleus. We can do this easily if we remember that for a traceless EFG tensor the sum of the diagonal elements is zero:

$$V_{xx} + V_{yy} + V_{zz} = 0$$

If the system is axially symmetric (i.e. $\eta = 0$), $V_{yy} = V_{xx}$, and it follows that if V_{zz} is zero then all three terms must be zero and the EFG vanishes. We thus need only calculate V_{zz} in an axially symmetric system. We take for the model in Fig. 4.6 a tripod of bonds to a nucleus N disposed regularly around the z axis with effective, equal charges q called q_1, q_2 and q_3 at a distance r from N.

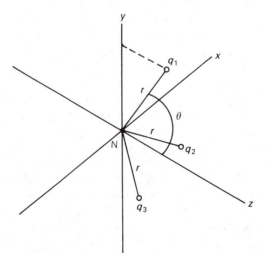

Fig. 4.6   Calculation of the EFG at N due to three equal charges q.

The z coordinate of each charge is $r \cos \theta$. According to equation (4.4) then

$$V_{zz} = 3qr^{-5}(3r^2 \cos^2 \theta - r^2) = 3qr^{-3}(3 \cos^2 \theta - 1)$$

In order that V_{zz} be zero, $(3 \cos^2 \theta - 1)$ must be zero, which requires that

$$\theta = 54.7356°$$

As we shall see, the quantity $(3 \cos^2 \theta - 1)$ appears in many expressions describing NMR phenomena, and the value of θ at which it becomes zero has become known as 'the magic angle'. It may seem astonishing that such a one-sided arrangement of charge should give zero EFG. If, however, we remember that the electric field is zero where the z axis intersects the plane defined by the three charges and is also zero at $-\infty$, we will see that it has to have a maximum at some point on the z axis, a point where its gradient is zero. We should also note that V_{zz} is zero for any number greater than three of equal regularly disposed charges that form the magic angle with the axis.

It has been customary to state as the necessary condition for low EFG a cubic arrangement of equal charges (or identical ligands) around a nucleus. This is correct, and indeed a nucleus at the centre of a regular tetrahedron, octahedron or cube has near-zero EFG. It is, however, too limiting a statement. The octahedron is in fact made up of two back-to-back tripods as depicted in Fig. 4.6 with all six charges equal and pairs of charges collinear with N. Equally, an octahedral complex in which there are two tripods of differing charge will still have zero EFG, i.e. an all-*cis* or *fac* structure NL_3M_3 where L and M are ligands. Such structures, while non-cubic, still have long relaxation times, and we can begin to see that long relaxation times for quadrupolar nuclei is a source of structural information.

A specific example is found with the molybdenum tricarbonyl arene complexes.

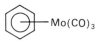

The nucleus ^{95}Mo has a significant quadrupole moment and its relaxation times vary typically from about 1 s in the regular tetrahedron $[MoO_4]^{2-}$ up to about 0.15 ms in less symmetric compounds. In the arene carbonyl complexes the times are near 0.07 s and are relatively long for a non-cubic symmetry. A to-scale sketch of the molecule is given in Fig. 4.7. The crystal structure of the mesitylene complex has been obtained and this shows that the three carbonyl ligands are orthogonal, which means that they each make the magic angle with the symmetry axis of the molecule. Following Fig. 4.6 then, the carbonyl ligands give zero net contribution to the EFG at the molybdenum. This means that the remaining bonding electrons to the arene ring must also produce a very small EFG at the metal. So the metal–arene bonds must lie on a conical surface with a half-apex angle equal to the magic angle also. This argument does not depend

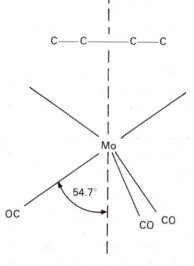

Fig. 4.7   A sketch of a molybdenum tricarbonyl arene complex. The carbonyl ligands·
are orthogonal and form half a regular octahedron. The lines pointing towards but
outside the arene ring have been drawn at the magic angle.

upon our knowing how many such bonds there may be; they have simply to be
disposed regularly on the conical surface. It is clear that such bonding orbitals
do not intersect the arene ring, that the conventional picture of overlap of metal
and arene orbitals is correct, and that the bonds are bent. It is also clear that
the electron density on the arene cannot influence the EFG and that this is
determined by the electron density nearest to the metal.

In general, however, many molecules either have some distortion of their
symmetry or their symmetry is non-cubic around the nucleus to be observed,
and the relaxation times and linewidths observed are very variable. Where
terminal atoms are concerned, the lines are often very broad and may not even
be detectable. This missing intensity problem has to be kept in mind when
working with quadrupolar nuclei, and it is often necessary to check the resonance
intensity against that of a standard sample in order to ensure that something
is not being missed. Even if the resonance is visible, its intensity will be incorrect
if its width is more than one-sixth that of the Fourier transform spectral width.
On the other hand, if the resonance is broad, but is resolved from other
resonances, then structural information is present in the spectrum. A few
examples of quadrupolar nuclei in different environments are given in Table 4.1
to illustrate the possibilities. Linewidths are quoted rather than relaxation times,

Table 4.1   *Linewidths of the resonances of some compounds of quadrupolar nuclei*

Nucleus	Molecule	Linewidth (Hz)	Comments
^{14}N	Me_4N^+	0.1	Regular tetrahedron, small $^{14}N-^1H$ coupling resolvable
^{14}N	Phenylammonium cation, $PhNH_3^+$	100	Non-regular tetrahedron
^{14}N	Methyl cyanide, MeCN	80	Terminal position
^{14}N	Methyl isocyanide, MeNC	0.26	Apparently linear and low EFG not possible. Requires annular electronic distribution around CN bond. See Fig. 4.6
^{27}Al	Hexaaquaaluminium cation, $Al(H_2O)_6^{3+}$	2	Regular octahedron, linewidth influenced by proton exchange
^{27}Al	Triisobutyl-aluminium	3990	Planar trigonal monomer with high EFG. Linewidth at ∞ dilution
^{14}N	Nitrate anion NO_3^-	3.7	Note contrast with previous example. The difference is explained by there being electron density above and below the plane of this trigonal ion
^{35}Cl	Perchlorate anion, ClO_4^-	1.2	Regular tetrahedron, aqueous solution
^{35}Cl	Phosphorus trichloride, PCl_3	7600	Terminal atom. Figure based on relaxation time measurement and is typical of many chlorides, e.g. CH_2Cl_2, $SiCl_4$.

since it is the linewidths that have been measured in most cases and these are proportional to R_{2Q} or $1/T_{2Q}$.

Solid-state NMR spectroscopy of quadrupolar nuclei or the zero-field technique of nuclear quadrupole resonance can give values for the interaction between the nuclear quadrupole and the EFG, and this is called the quadrupole coupling constant, χ, where in terms of equation (4.4)

$$\chi = eQV_{zz}/h = e^2qQ/h$$

If one can be certain that the EFG in solid and liquid are equal, and this likely to be so where the EFG is large and determined primarily by the bonding electrons rather than by long-distance effects in the solid, then it is possible to

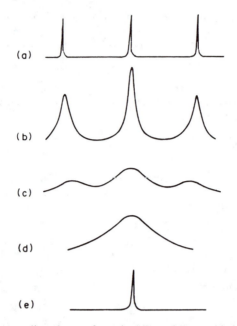

(a)

(b)

(c)

(d)

(e)

Fig. 4.8   The resonance line shape of a spin-1/2 nucleus coupled to a spin-1 nucleus having various rates of quadrupole relaxation. (a) T_1 long; (b) $T_1 \approx 8/(2\pi J)$; (c) $T_1 \approx 3/(2\pi J)$; (d) $T_1 \approx 1/(2\pi J)$; (e) T_1 very short. The intensities are not to scale. The broadened lines are weak since the total area is constant.

obtain correlation times from the relaxation times of quadrupolar nuclei in molecules in solution.

4.3.1   Spin–spin coupling to quadrupole relaxed nuclei

Longitudinal relaxation involves changes in spin orientations relative to B_0, and so we might except relaxation processes to modify the spin coupling patterns that we observe. In fact, the relaxation times of spin-1/2 nuclei are sufficiently long that relaxation normally has no effect on the analysis of the spectra, but the much shorter relaxation times of the quadrupolar nuclei do lead to considerable modification of the observed patterns.

If T_{1Q} is long, then the normal spin coupling effects are observed, such as the coupling of protons to both ^{10}B and ^{11}B shown in Fig. 3.8.

At the other extreme, if T_{1Q} is very short, the nuclear spins interchange energy and change orientation so rapidly that a coupled nucleus interacts with all possible spin states in a short time. It can distinguish only an average value of the interaction and a singlet resonance results. This explains, for instance, why the chlorinated hydrocarbons show no evidence of proton spin coupling to the chlorine nuclei ^{35}Cl and ^{37}Cl, both with $l = 3/2$. At intermediate relaxation rates, the coupling interaction is indeterminate and a broad line is observed. The line shapes calculated for the resonance of a spin-1/2 nucleus coupled to a quadrupole relaxed spin-1 nucleus such as ^{14}N are shown in Fig. 4.8. The shape of the spectrum observed depends upon the product $T_1 J$, where T_1 is the relaxation time of the quadrupole nucleus, since if the frequency defined by $1/(2\pi T_1)$ is comparable with the coupling constant in hertz, then the coupled nucleus cannot distinguish the separate spin states. The situation is equivalent to attempting to measure the frequency of a periodic wave by observing only a fraction of a cycle. The resonance of the quadrupolar nucleus will, of course, be split into a multiplet by the spin-1/2 nuclei, but each component line will be broadened by its relaxation.

A common example of lines broadened by coupling to quadrupolar nuclei is found in amino compounds. The protons on the nitrogen are usually observed as a broad singlet, a good example appearing in Fig. 3.18(b). It is important to remember in this case that the relaxation time of the amino proton is unaffected and can cause normal splitting in vicinal protons bonded to carbon. In contrast, the protons in the highly symmetric ammonium ion give narrow resonances because the nitrogen quadrupole relaxation is slow (Fig. 3.8).

Since quadrupole relaxation is sensitive to temperature and viscosity, the line shapes observed for coupled nuclei are altered by viscosity and temperature changes, an increase in temperature leading to *slower* relaxation and a better resolved multiplet. This fact is stressed, since on a first encounter it seems contrary to one's expectation. Alternatively, lowering the temperature increases the rate of quadrupole relaxation and, if the resonance of the coupled spin-1/2 nucleus was already broad, this may well narrow as the relaxation becomes too fast for any coupling to be exhibited. This technique can sometimes be used to advantage to simplify the spectra of spin-1/2 nuclei coupled to quadrupolar nuclei.

4.4   Spin rotation relaxation: detailed molecular motion

Theories of molecular motion differentiate two linked types of motion, which are given two correlation times τ_2 and τ_J. The time τ_2 is the orientational correlation time or the time between significant changes in orientation. This is clearly equivalent to the correlation time τ_c used here, which can be obtained from measurements of T_{1DD} or T_{1Q}, both of which measure the reorientation of internuclear vectors, which may or may not correspond to bonds. The time τ_J is the angular momentum correlation time or the time between significant changes in angular momentum. τ_2 and τ_J are not independent since, if the angular velocity is maintained for a longer time and τ_J is long, the orientation must change more rapidly and τ_2 is short. Alternatively, rapid random changes in angular velocity leave the orientation more or less unchanged. Thus τ_2 and τ_J are related by

$$\tau_2 \tau_J = A/T$$

where T is the absolute temperature and A is a constant, whose value depends upon the detailed model used to explain the motion. Understanding the details of the process thus requires measurement of both correlation times in order to determine A.

This molecular rotation has a further effect upon nuclear relaxation, which can be detected if the dipole–dipole mechanism is weak. If the rotation is particularly fast, the system of bonding electrons is subject to some displacement relative to the atomic nuclei, which gives rise to a small magnetic moment proportional to the angular momentum of the molecule. Changes in the direction of the magnetic moment provoke relaxation in the same way as with other processes, but with the difference, which is unique to spin rotation relaxation, that the faster the rotation, the longer the corresponding correlation time τ_J. Thus increasing the temperature increases the efficiency of this relaxation process. A study of the temperature dependence of the nuclear relaxation time can thus distinguish the spin rotation mechanism from the other mechanisms. The notation used by NMR spectroscopists describes the relaxation time as T_{1SR} and τ_J is then called τ_{SR}. T_{1SR} is significant for small molecules, or rapidly rotating side groups in larger molecules. The mechanism is particularly efficient in the gas phase, and thus resonances of gases are much broader than observed in solutions of the same molecules.

Two examples will illustrate these points. In the first, the relaxation time of the nitrogen nuclei is measured in liquid nitrogen under pressure and over a wide range of temperature. The two types of motion were separated by using $^{14}N_2$, which relaxes by the quadrupolar mechanism and gives T_{1Q} and so τ_c, and nitrogen enriched in the spin-1/2 nucleus ^{15}N, which, because of its small magnetogyric ratio, relaxes by the spin rotation mechanism and gives T_{1SR} and so τ_{SR}. The results are shown in Fig. 4.9, where the opposite temperature dependences are obvious. The second example concerns the relaxation behaviour of the quadrupolar nucleus ^9Be in the hydrated cation $Be(H_2O)_4^{2+}$. The quadrupole moment of this nucleus is small, so that the quadrupole relaxation mechanism does not necessarily dominate its relaxation. Interaction with the protons of the water ligands causes dipole–dipole relaxation, but this can be eliminated by using deuteriated water as solvent. The results are shown in Fig. 4.10, where it will be seen that the relaxation shows three different types of behaviour with change in temperature: T_1 increases with temperature at low temperatures, then has a maximum where there is little change and then decreases with further increase in temperature. Replacing H_2O by D_2O increases the relaxation time at low temperature but has no effect at high temperature. In the low-temperature region, then, for $Be(H_2O)_4^{2+}$ we have both quadrupolar and dipole–dipole relaxation, and for $Be(D_2O)_4^{2+}$ we have effectively only quadrupole relaxation, so that

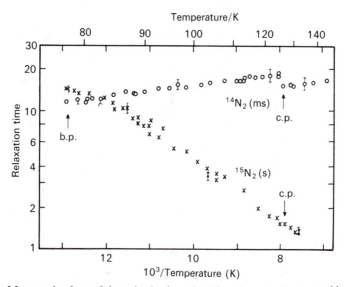

Fig. 4.9   Measured values of the spin–lattice relaxation time. T_1, for liquid $^{14}N_2$ (in ms) and liquid $^{15}N_2$ (in s) on the liquid–vapour coexistence line *versus* temperature (on a scale of $10^3\,K/T$, where T is the absolute temperature): c.p. indicates the critical point, and b.p. the boiling point at 1 atm. (From Powles *et al.*, *Mol. Phys.* **29**, 539; reprinted with permission.)

the measurements in the two solvents allow us to separate the two processes. Evidently, even for this tetrahedral molecule with low EFG, the quadrupolar mechanism is the most effective. At high temperature the changed temperature dependence indicates that the spin rotation mechanism now predominates. The various relaxation rates at 80°C are approximately $R_{1DD} = 7.7 \times 10^{-3}$, $R_{1Q} = 4.3 \times 10^{-2}$ and $R_{1SR} = 0.24\,s^{-1}$ respectively, the first two processes accounting for less than 18% of the overall relaxation rate.

4.5   Chemical shift anisotropy relaxation

We have already discussed in Chapter 2 how the chemical shift is a tensor quantity and varies as a molecule tumbles relative to the magnetic field direction. This averages to an isotropic value but nevertheless means that there is effectively a fluctuation in magnetic field strength at the nucleus that can also cause relaxation. The mechanism is not very efficient and depends upon

$$1/T_{1CSA} = (2/15)\gamma_I^2 B_0^2(\sigma_\parallel - \sigma_\perp)^2\tau_c \qquad (4.6)$$

where γ_I is the magnetogyric ratio of the nucleus whose relaxation time is required and the σ are defined in Chapter 2. This mechanism can be distinguished from others by the fact that it depends upon B_0^2. It is important

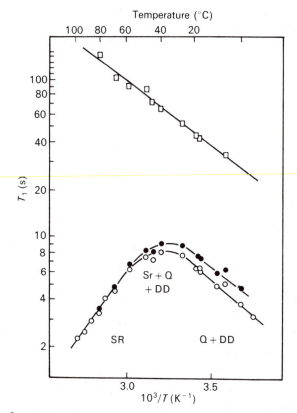

Fig. 4.10 The ^{9}Be spin–lattice relaxation time in 1 M aqueous Be(NO$_3$)$_2$ as a function of temperature: (○) H$_2$O solution; (●) D$_2$O solution; (□) T_{1DD} calculated from the difference between the rates of relaxation in both solvents. (From Wehrli (1976) *J. Magn. Reson.*, **23**, 181, with permission.)

for nuclei with a rather high screening anisotropy and high chemical shift ranges and at high magnetic fields. It leads to problems in ^{13}C spectroscopy in super-conducting magnets, and in specific instances at lower fields such as for ^{199}Hg in Me$_2$Hg where $\sigma_\parallel - \sigma_\perp$ is of the order of 7500 ppm.

An example has recently been reported that contrasts the relaxation behaviour of the spin-1/2 nucleus ^{195}Pt in the two anions Pt(CN)$_4^{2-}$ and Pt(CN)$_6^{2-}$. The former has square planar geometry and so a large chemical shift anisotropy (CSA), which has been measured to be $\sigma_\parallel - \sigma_\perp = -2500$ ppm in the solid state. The octahedral anion has no such anisotropy. The relaxation times were measured as functions of both temperature and of magnetic field. Two spectrometers were used, one operating at 4.70 T (^{195}Pt at 42.8 MHz) and one at 8.48 T (^{195}Pt at 77.04 MHz). The results are plotted in Fig. 4.11. Only one set are shown for Pt(CN)$_6^{2-}$ since the relaxation rates were the same at both

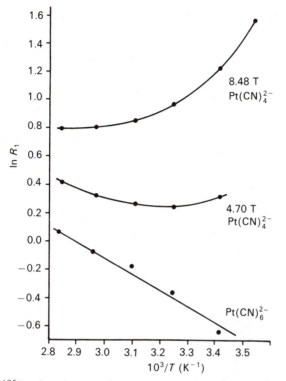

Fig. 4.11   The ^{195}Pt relaxation rates for aqueous solutions of $K_2Pt(CN)_4$ and $K_2Pt(CN)_6$ at $B_0 = 4.70$ and 8.48 T as functions of inverse temperature. The ^{195}Pt relaxation rates of $K_2Pt(CN)_6$ are independent of B_0. (From Wasylishen and Britten (1988) *Magn. Reson. Chem.*, **26**, 1076; copyright (1988) John Wiley and Sons Ltd, reprinted with permission.)

magnetic fields, the rate of motion being fast enough to ensure that the extreme narrowing condition was met. However, the rate of relaxation increases with temperature and so the mechanism of relaxation must be that of spin rotation, any dipolar contribution from the ^{14}N nuclei or from the solvent (D_2O) being small. The behaviour of the relaxation of $Pt(CN)_4^{2-}$ is quite different; the rate is increased at the higher magnetic field and the plots are curved, indicating the presence of two competing relaxation mechanisms, spin rotation at the highest temperatures and CSA at the lowest, with opposite temperature dependences.

4.6   Scalar relaxation

We have already mentioned that in many cases quadrupolar nuclei relax so rapidly that any effects of spin–spin coupling to other nuclei are completely lost. If, however, the rate of relaxation of the quadrupolar nucleus is very fast indeed, and if the coupling constant in the absence of the quadrupolar relaxation is significant, then the relaxation time of the coupled nucleus may be influenced by the quadrupolar nucleus. This is particularly true if the frequencies of the two nuclei are similar, when both T_1 and T_2 may have scalar contributions. In the case of the previous example, for instance, if the T_2 value of the ^{195}Pt equalled T_1, the linewidth would be around 1 Hz. In fact, it is 25 to 60 Hz, depending upon temperature, and this is due to coupling with the rapidly relaxing ^{14}N nuclei in the anion. In this case the frequency difference ^{195}Pt–^{14}N is too great for there to be any influence on T_1.

Questions

4.1. A spin-1/2 nucleus is relaxed by three different mechanisms and has a measured relaxation time of 1 s. The three contributions are spin rotation, whose characteristic time is 2.5 s, chemical shift anisotropy, with characteristic time 1.8 s and a long-range dipole–dipole interaction. Calculate the characteristic time of this latter contribution.

4.2. Given an isolated ^{13}C^{1}H$_2$ fragment in which the C–H distance is 109 pm and the H–H distance is 177 pm, use equations (4.2) and (4.3) to calculate the ratio of the rates of dipolar relaxation of the two nuclear species. The magnetogyric ratios should be taken as proportional to the nuclear frequencies given in Chapter 1.

4.3. It is pointed out in Table 4.1 that the electric field gradient at ^{14}N in the nitrate ion, NO_3^-, is almost zero, and that this is surprising. Use equation (4.4) to obtain an expression for the EFG at ^{14}N given that there are three charges q producing the field gradient and that these are situated in the N–O bonds at a distance r from the nitrogen atom. It will be found most convenient to choose the z axis as that perpendicular to the NO_3 plane and passing through the nitrogen atom, and to calculate V_{zz}. How far do we have to displace the charges from the NO_3 plane keeping r constant in order that the EFG will be zero at N? Does it make any difference as to which side we make the displacement?

5 The spectrometer

We have seen in the last three chapters that there are three principal parameters in an NMR spectrum, chemical shift, coupling and, less evidently, relaxation. The present chapter is devoted to showing in more detail how the spectrum is obtained and how these parameters may be derived from the data collected.

Chapter 1 discussed how a 90° B_1 pulse at the nuclear frequency can swing the magnetization from the direction of the magnetic field into the xy plane, and showed that this magnetization, which is now rotating at the nuclear frequency, can be detected in a suitable coil. In Chapter 4 we discussed how this magnetization is affected by a variety of relaxation processes; in particular, how T_2 leads to an exponential reduction with time of the xy magnetization. The spectrometer output thus consists of a signal at the nuclear frequency, which decays with time until it is no longer detectable. We need to know how this can be translated into typical spectral form. The equations governing the behaviour of the transverse and longitudinal magnetization M_{xy} and M_z and their return to equilibrium following a B_1 pulse are

$$(M_z)_t = (M_z)_\infty [1 - \exp - t/T_1)]$$

i.e. M_z increases from zero to its equilibrium value, and

$$(M_{xy})_t = (M_{xy})_0 \exp(-t/T_2)$$

i.e. the transverse magnetization falls from its maximum value (equal to M_z) to zero after sufficient time has elapsed. This behaviour is summarized diagrammatically in Fig. 4.2 above.

In general, the rate of decay of the spectrometer output is faster than predicted from the value of T_2 in a given system. This occurs because of inhomogeneities in the magnetic field throughout the sample due principally to imperfections in the magnet system, which mean that the nuclear frequencies are slightly different in different parts of the sample volume and this increases the rate at which the spins throughout the sample lose phase and coherence. We write that the apparent relaxation time, T_2^*, is equal to

$$1/T_2^* = 1/T_2 + 1/T_{\text{inhomo}}$$

where the last term is the decay in intensity due to the field inhomogeneities alone. Thus we cannot measure T_2 accurately from the rate of decay of the spectrometer output unless T_2 is so short that the inhomogeneity contribution is negligible. T_1 is only manifested in changes in intensity that occur following a series of closely spaced pulses, too close to allow M_z to come to equilibrium before the next pulse is applied. We will thus need to design specific experiments to measure the two relaxation times.

The nuclear frequency is high, several megahertz (MHz), and so is difficult to handle with normal data storage devices. For this reason it is necessary to reduce its frequency, and this is done using a device called a phase-sensitive detector, in which the nuclear frequency is compared with the frequency source used to generate B_1. The nuclear frequency thus is reduced to the difference between the two, which is zero to a few kilohertz, the intensity of each signal is preserved and, in addition, the phase of the signal is determined relative to the phase of B_1, the stimulating radiation. Knowledge of all three factors is essential in order that a spectrum can be extracted from the data, and knowledge of phase is particularly useful in some of the more complex experiments that we will encounter. It is useful to understand what the phase-sensitive detector does.

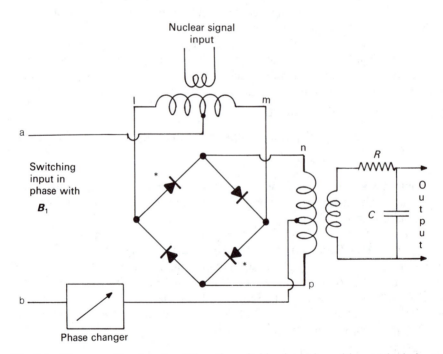

Fig. 5.1   Phase-sensitive detector. When the switching input is positive at 'a' relative to 'b', then the two diodes indicated by asterisks are in the conducting state. If the side 'l' of the input transformer is positive relative to 'm', the side 'n' of the output transformer is positive relative to side 'p'. If the polarity of the input changes, the other two diodes become conducting and the polarity of 'l' is transferred to 'p', thus reversing the output. RC is a filter to remove high-frequency components from the output.

5.1   The phase-sensitive detector

This device is essentially a switch driven by the B_1 signal and which reverses its polarity in step with B_1. A typical circuit is shown in Fig. 5.1. This consists of a ring of four diodes, each diode having the property that it is conducting if a voltage is applied in the direction of the arrow and is non-conducting if it is applied in the opposite sense. The B_1 signal thus renders one pair of diodes conducting during its positive half-cycle and the second pair conducting during the other, negative, half-cycle. The input nuclear signal, which is much smaller in intensity, is thus directed in one direction during the positive half-cycle and in the opposite sense during the other half-cycle. If the B_1 and nuclear frequencies are exactly the same, then the nuclear signal is rectified and a direct current output is obtained. If the two signals are in phase, then the output is positive; if they are 180° out of phase, the output is negative; and if they are of intermediate phase, then the output is reduced and is zero when the phase difference is 90°. The nature of the output is determined by the phase shifting device shown and which can be adjusted at will. If the nuclear and B_1 frequencies are not the

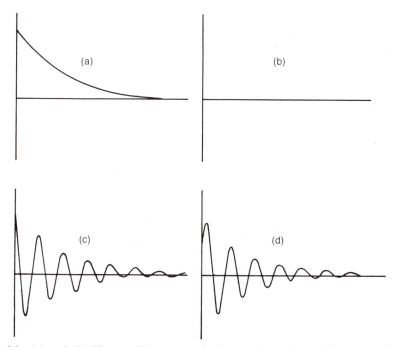

Fig. 5.2   (a) and (b) The possible range of outputs when the nuclei are exactly on resonance with B_1 and are in phase (a) or 90° out of phase (b). If the nuclear and B_1 frequencies are different, then the output is at the difference frequency, but starts on a maximum if they start in phase (c) or a minimum if they start 90° out of phase (d). Any intermediate phase is possible.

same, then their relative phase fluctuates with time and the output becomes a wave oscillating at the difference frequency. The phase information is still retained since the output will have some phase angle relative to B_1 immediately after the 90° pulse, i.e. at the start of the decaying output. The spectrometer output after phase-sensitive detection is thus oscillatory, reference to B_1 and with much reduced frequency. It decays with time characterized by T_2^* (Fig. 5.2).

5.2   Exponential relaxation and line shape

The T_2 relaxation process means that the nuclear frequency is not precisely defined and that the spins can be thought of as having a frequency distribution around their resonance frequency ω_0. A plot of this frequency distribution against frequency is the line shape $f(\omega)$ (Fig. 5.3). Individual nuclei have quite random frequencies within this range, but over a short enough time interval we can regard them as having this quite regular behaviour.

If we have an assembly of N nuclei, we can imagine that a small proportion n_k will have a particular angular velocity, ω_k. Each n_k is given by the line shape function

$$n_k \propto f(\omega_k - \omega_0)$$

which we can normalize by writing

$$n_k = N f(\omega_k - \omega_0)$$

i.e. we have chosen n_0 so that $\sum_k f(\omega_k - \omega_0) = 1$.

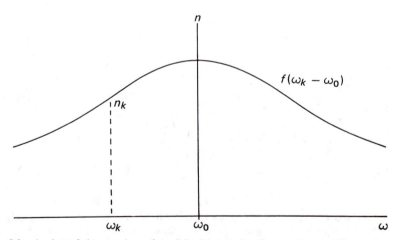

Fig. 5.3   A plot of the number of nuclei with angular frequency ω_k different from the resonant frequency ω_0 for a small time interval t. The function $n_k = f(\omega_k - \omega_0)$ is the line shape function.

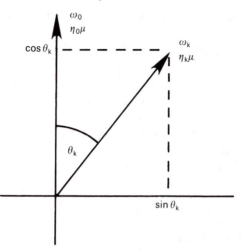

Fig. 5.4 How the total magnetic moment of the group of nuclei n_k with angular velocity ω_k is displaced an angle θ_k from the on-resonance nuclei time t after a $\boldsymbol{B}_1$ pulse has produced magnetization in the xy plane.

At a small time t after the $\boldsymbol{B}_1$ pulse, those nuclei with angular velocity ω_k will have moved an angle θ_k from those nuclei exactly at resonance, ω_0 (Fig. 5.4). We can resolve the magnetic moments of the n_k nuclei along the ω_0 direction and at right angles to it, and, following standard alternating current theory, we shall differentiate the normal component by the prefix i, where $i = \sqrt{(-1)}$.

We have that

$$\theta_k = (\omega_k - \omega_0)t$$

and the two components are

$$n_k \cos[(\omega_k - \omega_0)t] \qquad \text{and} \qquad in_k \sin[(\omega_k - \omega_0)t]$$

The next step is to replace the n_k by the line shape function. However, Fig. 5.2 indicates that these may not be the same for the two components, and so we shall denote them as two separate functions by the letters v and u. The total intensity of the two components is then proportional to

$$Nv(\omega_k - \omega_0)\cos[(\omega_k - \omega_0)t] \qquad \text{and} \qquad iNu(\omega_k - \omega_0)\sin[(\omega_k - \omega_0)t]$$

Summing over all possible values of k gives us the total intensity of each component. We normally would wish to compare this with the initial intensity at $t = 0$. This is proportional to the sum of all the nuclei, N. Thus N cancels from the formulae. The intensity of the normal component is, of course, only significant if $\omega_0 \neq \omega_{B_1}$. If N is large we can use the integral form of the equations, which, remembering that ω is the variable and t is constant, gives

$$\int_0^\infty v(\omega_k - \omega_0) \cos[(\omega_k - \omega_0)t]\,d\omega = e^{-t/T_2} \tag{5.1}$$

and

$$\int_0^\infty u(\omega_k - \omega_0) \sin[(\omega_k - \omega_0)t]\,d\omega = e^{-t/T_2} \tag{5.2}$$

It remains to find the form of the functions v and u. Inspection of tables of definite integrals will show that these are

$$u = \frac{(\omega_k - \omega_0)T_2^2}{1 + T_2^2(\omega_k - \omega_0)^2} \tag{5.3}$$

$$v = \frac{T_2}{1 + T_2^2(\omega_k - \omega_0)^2} \tag{5.4}$$

These two functions are plotted in Fig. 5.5; u is the dispersion mode and v is the absorption mode of the Lorentzian line shape. The absorption mode has its maximum at ω_0 and its intensity is proportional to the number of nuclei producing the signal. The dispersion mode is of zero intensity at resonance and of different sign above and below resonance. Spectra are displayed in the absorption mode, though, as we shall see, the dispersion mode has an important use in spectrometer locking systems.

The problem remains of how to obtain the plots of Fig. 5.5 from the actual spectrometer output of Fig. 5.2. This is known as the free induction decay (FID) and is stored digitally in computer memory where it can easily be mathematically processed. Time and frequency domains are related through the Fourier relationship

$$F(\omega) = \int_{-\infty}^{\infty} f(t)e^{-i\omega t}\,dt$$

This can be written as

$$F(\omega) = \int_{-\infty}^{\infty} f(t)[\cos(\omega t) - i\sin(\omega t)]\,dt \tag{5.5}$$

The Fourier transform of the output data thus contains two components, often called real (R) and imaginary (I), which correspond to the u and v components. The inverse transform is also possible, i.e.

$$f(t) = \frac{1}{2\pi} \int_{-\infty}^{\infty} F(\omega)e^{i\omega t}\,d\omega$$

and this should be compared with the relations (5.1) and (5.2) derived above.

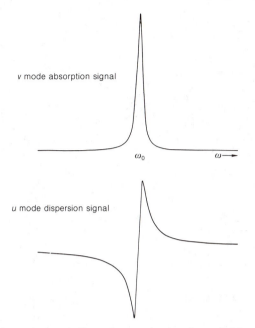

v mode absorption signal

ω_0 $\omega \longrightarrow$

u mode dispersion signal

Fig. 5.5   The absorption and dispersion mode signals available at the output of the phase-sensitive detector.

5.3   Time and frequency domains

The Fourier transform (FT) is a mathematical relationship that relates a function of time to one of frequency, i.e. the time and frequency domains. The output of an NMR spectrometer is a sinusoidal wave that decays with time, varies entirely as a function of time and so exists in the time domain. Its initial intensity is proportional to M_z and so to the number of nuclei giving rise to the signal. Its frequency is a measure of its chemical shift, and its rate of decay is related to T_2 and the quality of the magnetic field. Fourier transformation of this FID gives a function whose intensity varies as a function of frequency and is said to exist in the frequency domain. The parameters of the absorption curve contain all those of the FID: the position reproduces the frequency and so the chemical shift. Equation (5.4) can be used to predict the linewidth, which we define as the width at half-intensity. The half-intensity points occur when

$$1 + T_2^2(\omega_0 - \omega)^2 = 2$$

i.e. when

$$T_2^2(\omega_0 - \omega)^2 = 1$$

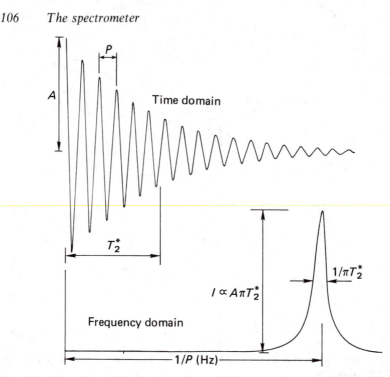

Fig. 5.6   The relationships between time and frequency domains. The period P of the FID gives the position of the line. The rate of decay T_2^* gives the linewidth and the initial amplitude A gives the line its area and therefore its intensity proportional to $A.\pi T_2^*$.

or

$$\omega_0 - \omega = \pm 1/T_2$$

The separation of the half-intensity points is then twice this in radians. It is, however, more usual to express linewidths in hertz (frequency $= \omega/2\pi$), giving

$$\nu_{1/2} = 1/\pi T_2 \tag{5.6}$$

where $\nu_{1/2}$ is the frequency separation of the half-height points. The linewidth thus gives us $1/\pi T_2^*$. The intensity of the absorption curve depends upon linewidth and, in fact, the product of linewidth and peak height is proportional, to the area under the curve which is proportional to the initial FID intensity and so to the nuclear concentration. These relationships are summarized in Fig. 5.6.

   In general, any function of time will have an equivalent representation in the frequency domain, and it is of interest to consider some examples, which are illustrated in Figs 5.7, 5.8 and 5.9.

1. An infinitely long wave of constant intensity is represented by a single

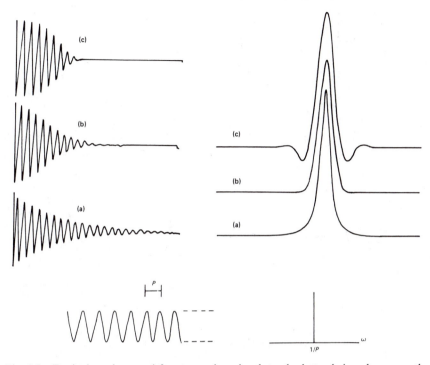

Fig. 5.7   Equivalent time- and frequency-domain plots, the latter being shown on the right of the figure. The lowest trace is a cosine wave of period P, which gives a single infinitely narrow line at frequency $1/P$. (a) An exponential decay $I_t = I_0 e^{-nt}$, which gives a Lorentzian line with sharp top but wide skirts. (b) A Gaussian decay $I_t = I_0 e^{-(nt)^2}$, giving a line with narrow skirts but a thickened top. (c) A super-Gaussian decay $I_t = I_0 e^{-(nt)^2}$, which introduces flanking waves at the base of the peak. (From Akitt (1978) *J. Magn. Reson.*, **32**, 311, with permission.)

monochromatic frequency. The precision with which we may measure this frequency depends upon how long we are willing to spend in its measurement.

2. Any distortion of this wave in the time domain results in the production of extra frequency components in the frequency domain. If the wave persists only for a limited time, then a packet of frequencies is produced, which gives the line definite width in the frequency domain. The line shape depends upon how the wave decays in the time domain. If the decay follows the law

$$I_t = I_0 e^{-(nt)^P} \tag{5.7}$$

then if $p = 1$, we have an exponential decay as illustrated in Figs 5.6 and 5.7(a). If $p = 2$, we have a Gaussian decay; and if $p = 4$, we have a super-Gaussian decay. These and their frequency-domain equivalents are shown in Fig. 5.7.

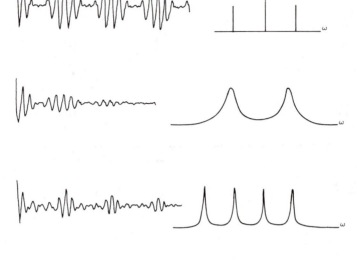

Fig. 5.8   Time- and frequency-domain equivalents of multiple-frequency responses containing two, four and six components. The time-domain signals are relatively simple because the components have been chosen to be of equal amplitude and regularly spaced. If the spacing and amplitudes are irregular, the time-domain signal becomes much more complex.

3. If several frequency components are present in the time-domain signal, then these interfere and produce a beat pattern such as shown in Fig. 5.8. Each component gives a line in the frequency domain with width determined by the rate of decay of the time-domain signal. Note that, when there are many component present, the time-domain signal starts to resemble a series of short pulses.

This introduction of new frequencies with distortion of the wave is a general phenomenon, and one way of stating the Fourier theorem is to say that any periodic function of time, however complex, may be synthesized from a suitable combination of pure cosine waves.

One waveform that we will use consistently in producing spectra is the rectangular pulse generating B_1. Usually a series of pulses is used to produce a series of responses, which are added together. A single rectangular pulse (Fig. 5.9) has a very rapid decay, which can be represented as following the law of equation

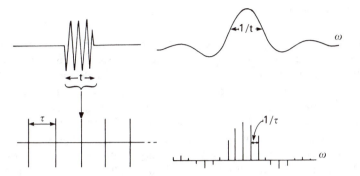

Fig. 5.9   The frequency-domain equivalents of short, rectangular pulses. The upper pair of traces show a single pulse of length t seconds whose frequency is spread out over a range of $1/t$ Hz. The lower pair of traces show a train of such pulses separated by an interval τ seconds. In this case the frequency-domain signal is not continuous, but consists of peaks of energy separated by $1/\tau$ Hz with the same spectral envelope as the single pulse.

(5.7) with the exponent p being very large. The flanking waves in the frequency domain are now much more pronounced, and the frequency intensity follows a sinc curve, where $\operatorname{sinc} X = (\sin X)/X$. The width of the central part of the response is the inverse of the length of the pulse and, since NMR spectrometers are designed to have $90°$ pulses of length generally less than $100\,\mu s$, the width of the frequency coverage of the $\boldsymbol{B}_1$ field is at least $10\,kHz$. Thus the use of short $\boldsymbol{B}_1$ pulses ensures that all the nuclei in a sample, whatever their chemical shift, are swung around $\boldsymbol{B}_1$ by an appropriate angle. Intensity distortion only occurs if the chemical shift range is very large. The same comments apply to a train of pulses, except that the frequency coverage is now discontinuous. This has little effect since the time between pulses will normally be longer than T_2^* and the spacing between the peaks of energy will be less than a linewidth. In all cases we will have present simultaneously in the output signals due to the nuclei in a sample, and the output will be complex and so require mathematical analysis to obtain the frequency-domain spectrum.

It is also possible to produce a frequency-selective pulse. The spectrometer output has to be much reduced so that a $90°$ pulse has to be very long. In such a case the frequency spread of the pulse will be only a few hertz and one can choose nuclei of a particular chemical shift to precess around $\boldsymbol{B}_1$ without affecting any other nuclei in the sample.

It should be clear from all this that there is an inverse relationship between time intervals in the time domain and spread of frequency in the frequency domain. The degree of resolution in the spectrum (frequency domain) is thus determined by the time for which the response is collected following the $\boldsymbol{B}_1$ pulse. The minimum linewidth is determined by T_2^*, but if the time of collection is shorter than T_2^* then this maximum degree of resolution cannot be attained.

5.4   The collection of data in the time domain

In order to carry out numerical processing of the output of an NMR spectrometer, we need to convert the analogue electrical output into digital information using an analogue–digital converter. The signal is sampled at regular intervals, the voltage registered converted into a binary number, and this number then stored sequentially in one of a series of computer memory locations, one location being a word of 16 or 20 bits (Fig. 5.10). The magnitude of the sampling interval is very important, and is often known as the dwell time, DW. The maximum frequency or minimum period that sampled information can represent is limited to that where alternate numbers are positive and negative – assuming a cosine wave input varying symmetrically about zero. The period of this wave is then $2(DW)$ and its frequency is $1/[2(DW)] = F$. A lower frequency, $F - f$, or a higher frequency, $F + f$, give the same pattern in memory, so that there is an ambiguity in the digital representation of the signals. This is avoided by ensuring that all the nuclear frequencies lie in the range 0 to F Hz. This is often known as the sweep width or as the spectral width. F is known generally as the Nyquist frequency.

Given that the spectrometer is highly stable, a series of isolated 90° pulses will each give an identical nuclear response, so that these can be added together in the computer memory to give a much stronger total response. Each signal is, however, accompanied by unwanted random noise; indeed, weak signals may not be visible because of the noise. While the noise intensity also increases as more responses are added, the noise is incoherent, i.e. its intensity at a given memory location is sometimes + and sometimes − , and so it adds up relatively slowly. A series of N FIDs when added together in this way has a signal-to-noise ratio that is $\sqrt{N}$ times better than that of a single FID. It is this feature that renders Fourier transform spectroscopy so useful for the less receptive nuclei, allowing a response to be collected from an apparently hopeless morass of noise (Fig. 5.11).

5.5   Memory size

The time during which a single FID may be collected is limited by the finite size of the computer memory. If there are M locations, then they will all have been traversed after $M(DW)$ seconds. Since, strictly, we should wait until $5T_1$ seconds have elapsed and the system is again at equilibrium before we apply the next pulse, we may also have to introduce a waiting time. Thus a standard FT experiment can be summarized by Fig. 5.12. The size of the memory does not affect the spectral width. After transformation of the accumulated FID, the dispersion and absorption parts of the spectrum each occupy $M/2$ locations of the same memory (this gives maximum utilization of expensive memory space), and the smallest frequency interval that can be detected is $2F/M$ Hz. Memory

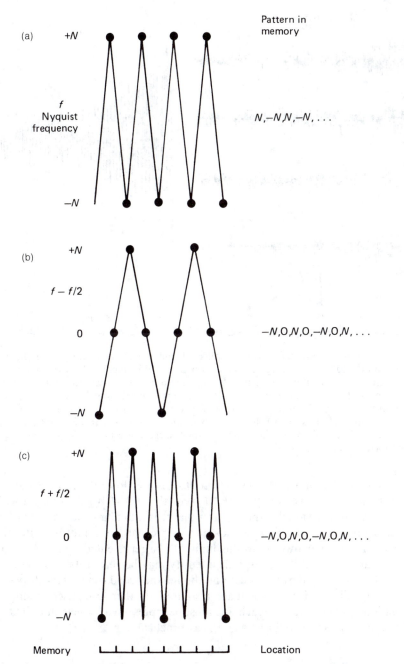

Fig. 5.10 (a) How a waveform at the Nyquist frequency, F, gives alternate positive and negative values of the number N corresponding to peak voltage. This assumes that the wave and the computer memory sweep are in a certain timing relationship. (b) A waveform of lower frequency, $F - F/2$, gives a different pattern. (c) A waveform of frequency higher than the Nyquist frequency by the same amount, $F + F/2$, gives exactly the same pattern.

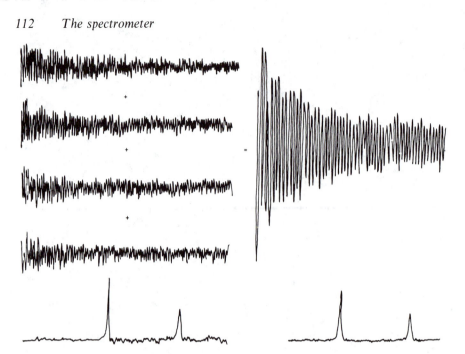

Fig. 5.11 Four separate FIDs were collected and stored in memory, and these are plotted at the top left of the figure. The FID becomes indistinguishable from noise about one-third of the way along and has only twice the noise intensity even right at the start. The sum of the four FIDs is shown at the top right, and the signal now starts with an intensity that is much larger than noise. The Fourier transforms of the sum and of a single initial FID are shown below. The theoretical improvement for four FIDs is two times, though, in this particular example, some large noise fluctuations seem to have been fortuitously suppressed.

size therefore determines the resolution. If lines are closer than $2F/M$ they can never be separated. We note that $2F/M$ is equal to $1/M(DW)$, the time for which we observe an individual response, and that it is a fundamental fact that if we observe a response for only t seconds then our resolution cannot be better than $1/t$ Hz. Any line narrower than this limit will be represented by a single point in the transform and its absorption intensity may be zero if its frequency falls between those defined by two locations, though the broad wings of the dispersion part will still be visible. Adequate resolution is therefore essential for full definition of the spectrum – each line must be represented by several points and so the memory size should be as large as practicable.

5.6   Rapid multiple pulsing

The waiting time needed to allow the spin system to come to equilibrium is a waste of our equipment since it is quiescent while waiting. It was quickly found

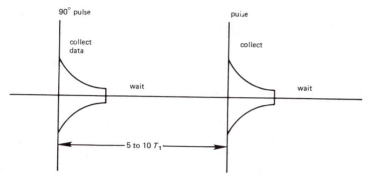

Fig. 5.12   The ideal Fourier transform experiment, which allows the spins to relax to equilibrium before successive pulses are applied. The data are collected until the memory is completely traversed and all memory locations contain data.

that the signal-to-noise ratio can be maximized in a given time if we adopt the procedure:

Pulse – acquire data – pulse – acquire ... and omit the waiting time. Thus we allow the computer acquisition time to determine the pulse repetition rate. This time will usually be much shorter than the ideal of Fig. 5.12, so that some transverse magnetization will still exist at the time of each pulse after the first. In fact, the later 90° pulses will tip the magnetization through the xy plane and the intensity will be reduced. The signal strength is then optimized by reducing the length of the pulse and so the pulse angle until a maximum steady-state output is achieved. The pulse lengths commonly found correct for this type of data collection are in the range of 40° to 30°, depending on nuclear isotope observed, chemical shift range, relaxation time T_1 and memory size. Now, the relaxation times of nuclei situated in different parts of the molecule will usually be different, so that the optimum pulse length will be different for each. This means that there will be a distortion of the intensity of the signal from each type of nucleus, which fortunately is not too grave a disadvantage in many cases since we know a molecule is made up of integral numbers of each type of atom. If precise quantitative data are essential, then the pulse sequence of Fig. 5.12 must be used.

5.7   Manipulation of collected data

FT NMR involves the collection of data in a computer memory. Now a computer is an infinitely variable instrument; it will do anything that you wish to the data that it contains, provided that you possess a suitable program. The minicomputers used for NMR spectroscopy have a section of memory containing programs as well as memory used simply for data accumulation, and many have in addition a backing store in the form of a disk or magnetic tape. Thus a set of data that may have taken several hours to acquire can be stored in permanent form and recalled to allow a variety of mathematical processes to be carried out with the object of improving the final spectrum. The technique most commonly used in the time domain is to multiply the data progressively by an exponentially decaying function, $k = e^{-n/K}$, where the multiplying factor k to use on the number stored in memory location n decreases from a value of unity for the first location. The constant K is a pseudo time constant related to the size of the data store to be treated and to the rate at which the data are to be attenuated as n increases. The process is illustrated in Fig. 5.13. This treatment increases the rate of decay of the FID and so broadens the lines obtained after Fourier transformation. This loss of resolution is normally acceptable because the signal-to-noise ratio is significantly improved. This comes about because the signal and noise have different shapes in the time domain; the signal is intense at first but decreases with time and may be undetectable towards the end of the FID, whereas the noise has no decay in intensity with time. The exponential multiplication thus decreases the noise much more than it does the signal. The effect of the process when applied to a noisy spectrum is shown in Fig. 5.14. It allows the signal-to-noise ratio to be increased in a time that is very much shorter than that needed to accumulate sufficient extra data. The amount of noise that is accumulated with a spectrum is determined by the rate at which data are accumulated, and there is relatively little flexibility in spectrometer setting possible. Thus, after a given time of accumulation, one has a certain signal-to-noise ratio, and this can be improved either by continuing with accumulation of data or by exponential multiplication

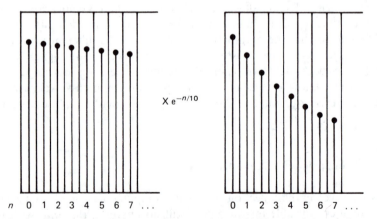

$X e^{-n/10}$

n 0 1 2 3 4 5 6 7 ...   0 1 2 3 4 5 6 7 ...

Fig. 5.13   An illustration of the way exponential multiplication is carried out in a computer. The boxes represent memory locations where numbers are stored, represented by the sticks. The slowly decaying function on the left is multiplied successively by $e^{-0.1}, e^{-0.2}, e^{-0.3}, \ldots$, for $n = 1,2,3,\ldots$, to give a more rapidly decaying function. Data collection is started in location 0, and n and time increase together after the pulse.

if the loss in resolution is acceptable. If it is decided to continue accumulation of data, it is necessary to continue for at least as long as has already been done to obtain a significant improvement. This is a simple rule of thumb and is reasonable in view of the square-root law for signal-to-noise improvement, which predicts a 40% improvement if the number of accumulations is doubled.

Multiplication of the data in the time domain by an exponentially decreasing function broadens the lines and reduces the noise level. It will come as no surprise that multiplication by an exponentially increasing function, $k = 1 + e^{n/K}$, has the opposite effect and, in principle, increases the noise level but reduces the linewidth. Such multiplication is known as resolution enhancement and is much used to reveal hidden details in spectra. It has the disadvantage that, as k becomes very large, the noise towards the end of the data is much multiplied and memory overfill occurs so that extreme spectral distortion is produced. This is avoided by tailing off the rise in k and even reducing its value again towards the end of the data. A commonly used function is the sine-bell function, which looks like two back-to-back rising exponential functions with a rounded maximum. The presence of noise limits what can be done in the way of resolution enhancement and such treatment requires the acquisition of good, noise-free data. Two examples are given in Figs 5.15 and 5.16 for spin-3/2 and spin-1/2 nuclei.

Figure 5.15 shows the ^{11}B spectrum of the borane B_5H_9 in which the influence of the protons has been removed by double irradiation (see later for an explanation of this). The four basal boron atoms are spin-coupled to the apical boron and are thus a quartet, whereas the latter is a broad, unresolved line, which we will not discuss further. The FID is shown in (a). Some 18.8% of the apical boron atoms in the molecules will in fact be the minor isotope ^{10}B with spin 3, so that the basal atoms in these molecules will be a septet, but this multiplet is obscured by the quartet from the remainder of the molecules; see spectrum (b). If the FID is multiplied by the limited rising exponential shown in (c), this produces a FID with a much longer apparent T_2 but also with rather a lot of noise in the tail. Suitable attenuation of this part takes out much of the noise without much affecting the early part of the FID; see (e). The resulting spectrum (f) exhibits three of the spin satellite lines, with the other four concealed within the skirts of the two central lines of the quartet. Spectrum (f) should be compared with the natural spectrum (g) where the quartet has been expanded to the same scale.

A second example is the determination of the long-range coupling between the protons in 2-chlorobenzaldehyde, a study of which permits the study of the barriers to rotation of the formyl group. Figure 5.16 shows two of the lines due to the proton H4 obtained with an instrumental resolution better than 0.05 Hz (A). Structure is visible in one of the lines, and after resolution enhancement the small coupling $^6J(f-4)$ is seen with a value of about 0.024 Hz. The variation of this very small coupling between the formyl proton and ring proton H4 with temperature enables the formyl group rotation to be studied. The wiggles in

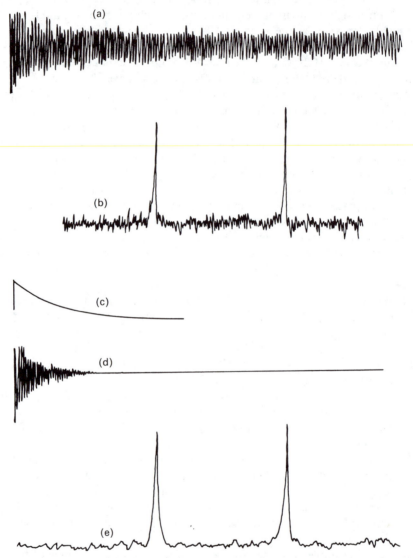

Fig. 5.14 The effect of the exponential multiplication process. (a) A noisy FID, where the signal quickly disappears below the noise, and (b) its transform. This FID is multiplied by the falling exponential (c) to give (d), where the most noisy part of the FID is much reduced, as is the noise in the transform (e), which is about half the intensity of that in (b) and smoothed. The decay rate of (c) was chosen to approximate that in the signal and is known then as a matched filter.

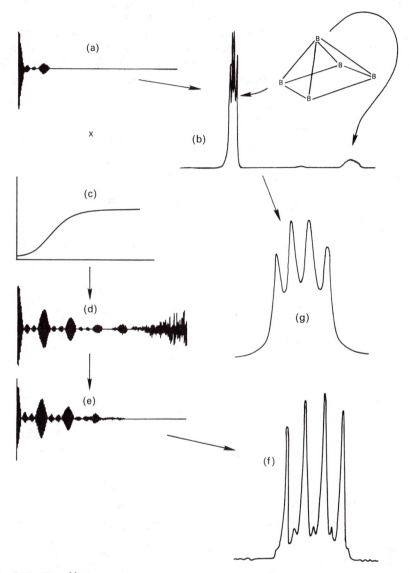

Fig. 5.15 The ^{11}B spectrum of B_5H_9 with proton double irradiation showing only boron–boron coupling. (From Akitt (1978) *J. Magn. Reson.*, **32**, 311, with permission.)

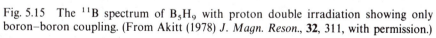

the baseline in thoth these spectra arise because of the rapid truncation of the data that occurs in the tail of the FID with the application of the noise-reducing function and which introduces a super-Gaussian element into the spectrum (cf. Fig. 5.7(c)). For this reason, resolution enhancement is also known as a Lorentzian–Gaussian transformation. The technique thus has to be used with care, since these wiggles can in extreme cases be mistaken for hidden peaks.

5.8   The Fourier transform

The production of the frequency-domain spectrum Equation (5.5) for the relationship between time and frequency domains is

$$F(\omega) = \int_{-\infty}^{\infty} f(t)[\cos(\omega t) - i\sin(\omega t)]\,dt$$

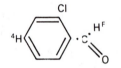

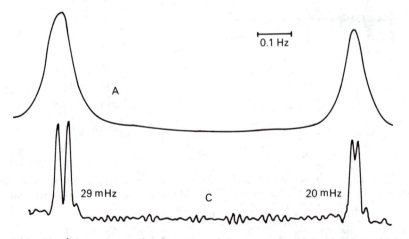

Fig. 5.16   The ^{1}H spectrum of a fragment of the spectrum of 2-chlorobenzaldehyde showing how resolution enhancement can reveal the very small coupling between the formyl proton and ring proton H4. (From Laatikainen *et al.* (1990) *Magn. Reson. Chem.*, **28**, 939–46; copyright (1990) John Wiley and Sons Ltd, reprinted with permission.)

It can be shown that it is valid to replace the continuous functions by discrete ones such as we collect in a computer, i.e. we can replace the integral symbols by summations and can carry out valid transforms in a computer. The process of Fourier transformation can be thought of as carrying out a series of multiplications of the time-domain data by sine and cosine functions that are stepped along the data. The results of this series of multiplications are summed and, if there is present in the data a wave of the same period, then there is an output; otherwise there is none. The use of both sine and cosine functions means that intensity and phase information are retained. The process is repeated for all possible frequencies up to the Nyquist frequency, so that all the signals present in the data may be detected. This would obviously be a very time-consuming process, even in a modern computer, and would require a vast amount of extra computer memory in order to store all the intermediate data needed. Fortunately, Cooley and Tukey have shown that, provided all the data are contained in a memory with number of locations equal to a power of 2 (e.g. 1024, 2048, 4096,..., etc., or, in computer jargon, 1 K, 2 K, 4 K,..., etc. locations, then the process can be factorized in such a way as to remove many redundant multiplications and can be carried out in the memory containing the data with only a few extra locations being needed to store numbers temporarily. The process is thus fast and economical of memory space. The existence of the Cooley–Tookey algorithm and of minicomputers were both necessary before FT NMR could become a commercial possibility. In fact, the current generation of dedicated computers are so advanced that they can perform a transform on accumulated data in another part of memory in order to observe how the accumulation is proceeding while at the same time more data are being added to the FID.

The sine and cosine multiplications produce, in effect, the two components of the nuclear magnetization in phase with and normal to B_1, and these are kept separate in the memory. The final result of the transform is thus two spectra, which should in principle be the absorption and dispersion spectra. This assumes that the B_1 signal fed to the phase-sensitive detector is exactly in phase with B_1, a condition that it is very hard to realize in practice. Thus the two sets of spectra contain a mixture of dispersion and absorption. In addition, we cannot be sure that we have started to collect data immediately after the B_1 pulse has ended; indeed, to avoid breakthrough, we may have purposely delayed the collection of data, commonly by one dwell time. This means that the starting point for each frequency component differs since each type of nuclear magnet will have precessed by a different amount by the time data collection starts, their phases are different and the degree of admixture of the two spectra varies across the spectral width. The two components thus have to be separated, a procedure known as phase correction, which is an essential part of FT spectroscopy. The two halves of the sets of spectra are weighted and mixed according to the formulae below:

$$\text{new disp} = (\text{old disp})\cos\theta - (\text{old abs})\sin\theta$$
$$\text{new abs} = (\text{old abs})\cos\theta + (\text{old disp})\sin\theta$$

where 'abs' means one half of the data and 'disp' the other. The multiplication is carried out point by point through the data and θ is adjusted until one line is correctly phased, i.e. purely absorption in one half of memory and purely dispersion in the other half. This has to be done by trial and error, monitoring the progress of the correction on the computer memory display screen. If not all the lines are then correctly phased, it is necessary to repeat the correction but allow θ to vary linearly as a function of position in memory. It is best to arrange that $\theta = 0$ for the correctly phased line and allow it to vary to either side. This second process is then continued until all the lines are correctly phased. Attempts have been made to write automatic programs to carry out this rather tedious process, but these have never been universally successful, and the two-stage process always has to be carried out by manual trial and error. The process is illustrated in Fig. 5.17.

Sometimes it happens that it is not possible to make a perfect phase correction. This may arise because the behaviour of the spin system is affected by the way the spectrum has been obtained – we have already mentioned that too rapid pulsing if spin–spin coupling is present may result in distortion of line intensities, and in such a case phase anomalies also occur. If we have not arranged that all the nuclear resonances occur within the sweep width covered by the computer, then those which are above the Nyquist frequency are discontinuously out of phase with the rest. Similarly, resonances on either side of the B_1 carrier frequency will contribute to the FID, though with different phase. It is thus important to ensure that all resonances occur on one side or the other of the carrier. If this is not done, then those resonances outside the range chosen will be reflected into the spectrum through the 'mirrors' formed by the limits of the computer sweep. This is called folding or aliasing. The chemical shifts and phases will be wrong, and indeed the phase anomalies assist in identifying when a spectrum is folded. Provision is made to vary the B_1 frequency over a range so that it can be set to bring a given sample into the computer window. Thus we ensure that the two folding ambiguities have been avoided by observing how the phase and shift anomalies vary with sweep width and B_1 frequency. Spectra illustrating both sorts of folding are shown in Fig. 5.18.

Modern systems use a more sophisticated detection system called quadrature detection, which avoids the folding around the B_1 carrier frequency and reduces the noise level by a factor of $\sqrt{2}$. The spectral width still needs to be adequate if folding around the Nyquist limit is to be avoided.

5.9   Manipulations in the frequency domain

Once we have the fully phase-corrected spectrum in memory, we now have to extract from it the information that we require. We will also need to make some

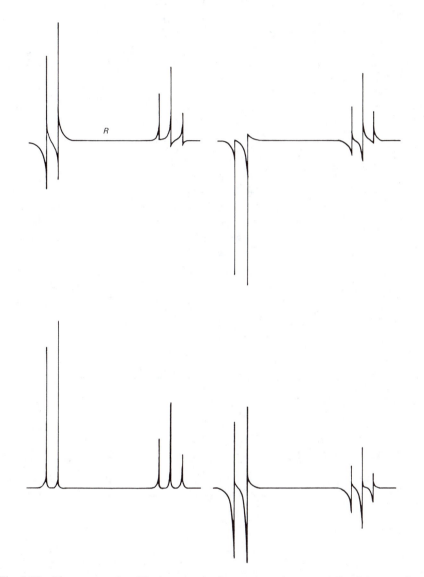

Fig. 5.17 Phase correction. The upper pair of spectra are the two results of the transform operation, R and I, with scrambled phase of the lines of the AB_2 spectrum. A two-parameter phase correction gives the purely absorption spectrum in one half and the purely dispersion spectrum in the other.

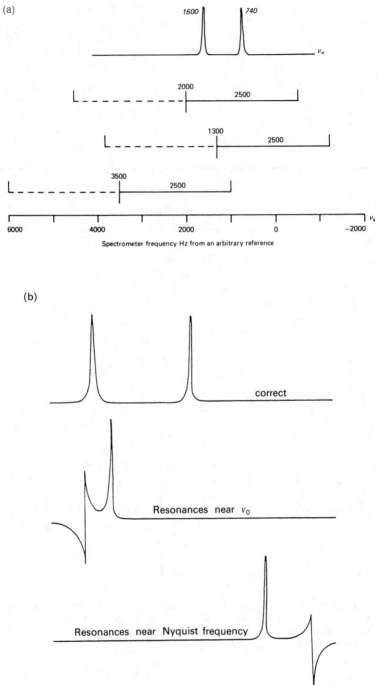

(a)

1600   740

ν_n

2000

2500

1300

2500

3500

2500

6000   4000   2000   0   −2000 ν_s

Spectrometer frequency Hz from an arbitrary reference

(b)

correct

Resonances near ν_0

Resonances near Nyquist frequency

sort of permanent record. Again, the computer allows us to carry out a variety of processes.

5.9.1   Visualization and plotting data

The contents of the data memory, whether the data are time or frequency domain, can be displayed continuously on an oscilloscope screen connected to the computer. The X axis is swept through the data memory and the numbers stored there are transferred sequentially to the Y axis. The result is a steady display of FID or spectrum, which allows the various manipulations to be monitored. A more permanent record is provided by plotting the data on a chart using an XY recorder and which operates in much the same way as the oscilloscope, though, of course, much more slowly. The recorder draws the graphic display on the oscilloscope screen. Precalibrated paper may be used, but the newer recorders are extremely versatile. Different sets of spectral data can be drawn in different colours together with expansions or integrals (see below), and they can also reproduce text and so calibrate the paper, give spot chemical shifts for individual lines and numerical integral values.

5.9.2   Integration

The initial intensity of a FID is proportional to the number of nuclei contributing to that signal and this transforms as the area of the Lorentzian absorption. An integral of the spectrum (we always, of course, refer to the absorption spectrum) then will tell us how many nuclei contribute to a given line and can give us invaluable quantitive data about a molecule – in the absence of relaxation effects discussed in Chapter 4. The integration is carried out simply by adding the numbers in successive memory locations. Where there is no resonance, this sum will remain constant, and will increase in the region of any resonance. The integral then forms a series of steps, rising at intervals, with each rise

---

Fig. 5.18   Folding of resonances around the ends of the computer memory 'window'. The spectrum consists of two resonances at 1600 and 740 Hz on an arbitrary scale related to the spectrometer frequency (13.8 MHz in this case for the ^2H nucleus). To set up correctly, we place the spectrometer frequency to the high-frequency side of both resonances, 2000 Hz in this case, and obtain an easily phase-correctable doublet, provided that the spectral width is large enough to accommodate the doublet; 2500 Hz being chosen in this case. If we reduce the spectrometer frequency to 1300 Hz, between the two lines, they appear at 300 and 440 Hz, close to the low-frequency end of the scan, conventionally on the left. The phase of one line is now altered. We obtain the same effect if we alter the frequency so that the Nyquist frequency is between the lines. Here they appear at 600 and 260 Hz. The phase difference between the normal and folded lines depends upon the phase setting of the B_1 feed to the phase-sensitive detector. A normal spectrum is obtained if v_s is set to the low-frequency side of the doublet but the lines are now reversed.

corresponding to a resonance (Fig. 5.19). For the plot to be vertical and horizontal as shown, it is necessary that the baseline is at zero intensity. Means are thus provided to correct the baseline to give the most acceptable integral.

5.9.3   Expansion

The typical oscilloscope screen is rather small to be able to observe much detail, and means are provided to enable a part of the spectrum to be selected and expanded to fill the screen to assist in the various operations needed to improve a spectrum and, for instance, show up small splittings.

5.9.4   Conserving data

Data can also be recorded on some form of magnetic storage system, at least for a limited period. It is usual to store the FID, which can then later be recalled and reprocessed to, say, show up some unexpected feature.

5.10   Continuous-wave (CW) spectroscopy

Virtually all NMR spectrometers used for chemical work up to about 1970 were CW spectrometers and FT spectroscopy has developed since that time because of the pressing need to obtain ^{13}C spectra routinely at natural abundance and aided by the technical developments mentioned above. The technique in principle uses a weak, infinitely long, infinitely selective pulse – a continuous wave. This can only cause precession of the spin vectors at one point in the spectrum and so the B_1 frequency has to be swept slowly through the spectral width and the M_{xy} magnetization detected and recorded directly as the absorption mode spectrum on an XY recorder whose X axis is swept synchronously with the B_1 frequency. No computer is required, so that the system is simpler and cheaper than a FT system, though its sensitivity is much less, principally because so much of the sweep time is spent searching areas where there are no resonances.

The level of the B_1 field is set so as to avoid tipping the nuclear vectors too far and is determined by the relaxation time T_1 and the time taken to traverse the lines and hence the sweep rate. The method is applied today only to the most receptive nuclei (^{1}H, ^{19}F particularly). It has certain advantages over the FT method and only a small portion of the total spectrum need be examined so that dynamic range is much less of a problem. Neither is the resolution limited by the memory size of the computer. It has the disadvantage, though, that, since the B_1 transmitter is on all the time, this interferes with the nuclear signal and adds noise and baseline instability. These are minimized by modulating the magnetic field by passing an a.c. current through coils in the magnet gap around the sample. If we choose a frequency of 4000 Hz, then the

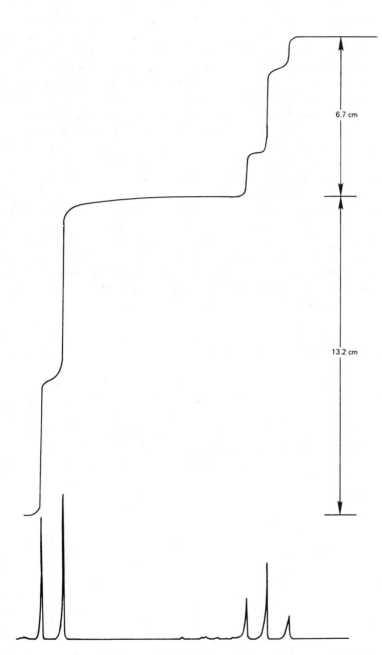

6.7 cm

13.2 cm

Fig. 5.19 A spectrum and its integral. The heights of the two steps are closely in the ratio required for an AB_2 spectrum, 2:1.

nuclear frequencies are modulated at this frequency and the signal can be carried by the resulting sidebands via a capacitor that blocks the low-frequency variations causing instability. The sidebands each consist of a full spectrum and the modulation frequency has to be larger than the spectral width.

The frequency sweep has one consequence that is not observed in FT spectrometers. When B_1 transverse a resonance, it creates magnetization M_{xy} precessing at the nuclear frequency, which can only become of significant intensity at resonance, but which will retain significant intensity for a time of T_2^* afterwards. The B_1 frequency changes continuously during this time and diverges from the nuclear frequency, and so beats with it via the phase-sensitive detector as the two go in and out of phase. The resonance is thus followed by a transient wiggle whose frequency of oscillation increases as the B_1 and nuclear frequencies diverge. The decay constant of the wiggles is related to T_2, though it is shorter than this because the increasing frequency is progressively attenuated by the bandpass filters used to minimize random noise. Wiggles are nevertheless a very useful phenomenon and are used universally to set up and maximize the homogeneity of the magnetic field by observing a sharp resonance under conditions of repetitive sweep and adjusting the shim currents until a prolonged smooth decay of wiggles is obtained.

5.11   Time-sharing spectrometers

These devices are in a way a compromise between FT and CW spectrometers. The transmitter output is in the form of weak, fairly long pulses at a repetition rate of, say, 4 kHz, and the receiver is on only when the transmitter is off (Fig. 5.20). This gives, effectively, CW operation without the baseline instabilities. There is no need to modulate the magnetic field since the transmitter pulse modulation provides the sideband frequencies that can carry the signal. This

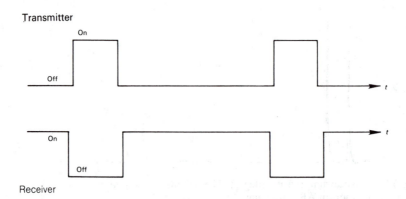

Fig. 5.20 Receiver/transmitter time sharing. The receiver is switched off just before the transmitter is turned on, and remains off for a little while after the transmitter is turned off.

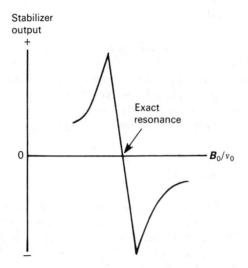

Fig. 5.21 A resonance in the dispersion mode can be used to stabilize an NMR spectrometer. At exact resonance, the output is zero; but if drift occurs, it becomes either positive or negative. The output can be used to provide a correction to the magnetic-field, which reverses the drift until the exact resonance position is regained.

means that the system is compatible with an FT device, so that we can have two spectrometers associated with one magnet. Thus these devices, which are relatively simple, are used in conjunction with FT spectrometers to provide a final stage of stabilization using a nuclear resonance in the sample other than the one to be observed – often the deuterium in a deuteriated solvent. The device is used in two ways. A small triangular repetitive sweep is applied temporarily to the magnetic field and the deuterium is observed in the absorption mode as its resonance is traversed repeatedly. We observe a signal followed by wiggles, and can use this to set up the magnetic field homogeneity. We then switch off the field sweep and turn our attention to the dispersion signal produced by the solvent. This has the property that, at resonance, the output is zero, but if drift of field or frequency occurs, either a negative or a positive output is obtained, the sign depending upon the direction of drift, and we can use this output to provide a correction voltage to alter the magnetic field until the output is again zero (Fig. 5.21). Thus the frequency and field are locked together indefinitely and we can in principle accumulate as many FIDs in memory as we wish, 100 000 being quite feasible, though the number is always kept as small as possible since there is very often a queue of people waiting to use any given spectrometer. It is, of course, necessary that the frequency of the lock device is derived from the FT drive frequency, otherwise they could vary independently. The lock then keeps the ratio B_0/ν_0 constant and is known as a field–frequency lock.

5.12   A modern system

The various parts of a complete FT spectrometer system are shown on the block diagram of Fig. 5.22. This summarizes much of what has been said in this chapter, but also includes a double-resonance device that we will need in Chapter 7. This is a third transmitter, which can provide power at the frequency of other nuclear isotopes in the sample, pulsed or continuous, and so affect their behaviour and modify the response of the nuclei observed.

5.13   Measurement of T_1 and T_2

Different experiments are needed to measure T_1 and T_2 accurately, though, if they are equal, then measurement of T_1 alone usually suffices. However, in many systems $T_2 < T_1$ and both need to be measured. In addition, an understanding of the method used to measure T_2 will prove helpful when we come to discuss two-dimensional NMR.

5.13.1   Measurement of T_1 by population inversion

This is the most popular of several available methods. We have already seen how a maximum NMR signal is obtained after a 90° B_1 pulse. The same is obtained after a 270° pulse, except that the spins have been inverted relative to the 90° pulse and the output is 180° out of phase with that after the shorter pulse. If we arrange our computer to give us a positive-going absorption peak from the FID following a 90° pulse, then after a 270° pulse we will get a negative peak. If, instead, we use a 180° pulse, we create no M_{xy} but turn the excess low-energy spins into the high-energy state, $-M_z$. They will relax to their normal state with characteristic time T_1 and the magnetization will change from $-M_z$ through zero to $+M_z$ (Fig. 5.23). This, of course, produces no detectable effects. However, if, at some time τ (do not confuse this with τ_c) after the 180° pulse, we apply a second pulse of 90°, we will create magnetization M_{xy} equal in magnitude to M_z at that instant. The spectrum that results from processing the FID will be negative-going if M_z is still negative (as if part of the magnetization had undergone a 270° pulse) and positive-going if M_z has passed through zero (Fig. 5.24). A series of spectra are obtained in this way for a number of different values of τ and the intensity of the resulting peaks plotted as a function of τ, so allowing us to extract a value for T_1 (Fig. 5.25). Sufficient time must elapse between each 180°/90° pair of pulses to allow M_z to relax fully to its equilibrium value or incorrect results will be obtaind. Usually a waiting period of five to ten times T_1 is used. If T_1 is very long, then the experiment can be very time-consuming, but other pulse sequences have been worked out that will allow the total time to be reduced.

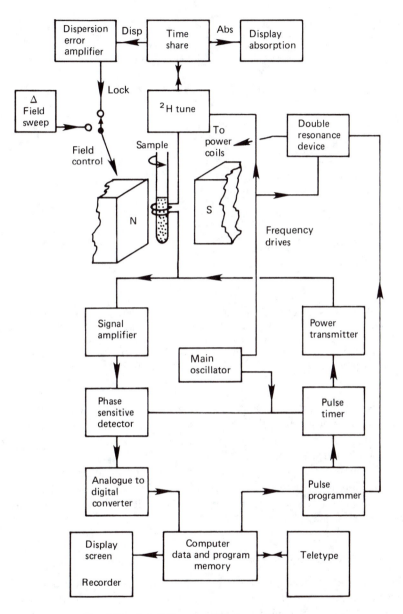

Fig. 5.22   A full Fourier transform (FT) system.

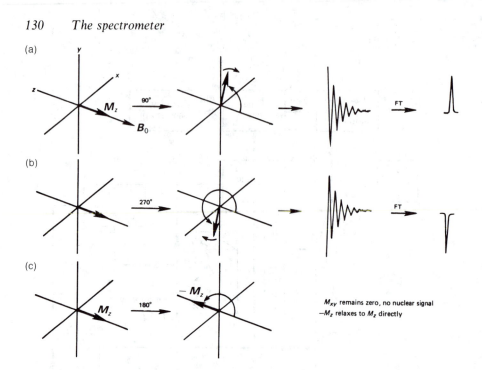

Fig. 5.23   (a) Production of a nuclear response using a 90° pulse. (b) A similar response is obtained after a 270° pulse, but 180° out of phase, giving an inverted transform. (c) A 180° pulse gives no xy magnetization but places M_z in a non-equilibrium position opposing the field. This magnetization relaxes back to its equilibrium value and no transverse magnetization is produced at any time throughout this process.

If there are several resonances in our spectrum, each relaxes at its own rate and it is possible with this method to measure the relaxation times of all the nuclei of the same isotope in a molecule and to obtain detailed information about molecular motion.

If the nuclei are spin-coupled then the apparent relaxation times may not be simply related to the real ones. Carbon-13 relaxation times are thus measured with the protons decoupled by double irradiation (see later) and this gives satisfactory values of T_1.

5.13.2   Measurement of T_2

The contribution of magnetic field inhomogeneity to T_2 will usually be of the order of 3.0 to 0.3 s for a high-resolution magnet. Thus, provided T_2 is less than about 0.003 s, it can be measured directly either from the linewidth or from the rate of decay of the FID. In the latter case we have to suppose that there is only a single resonance.

Accurate measurement of T_2 is made using a spin-echo experiment. This

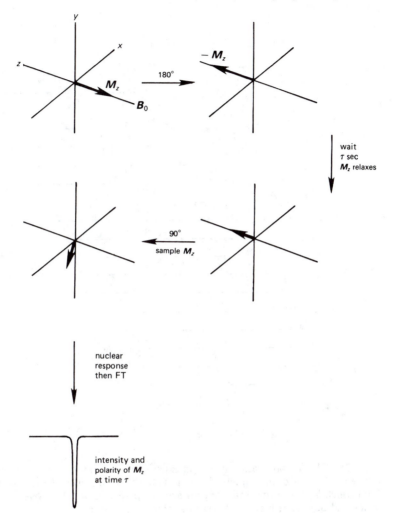

Fig. 5.24   How the pulse sequence $180°$ – wait τ–$90°$ can be used to perturb M_z and follow what then happens to M_z.

depends upon the rate at which the FID decays, being faster than the rate of longitudinal relaxation R_1. The decay is due to the T_2^* mechanism, which has two components. The loss of phase coherence between the spins in the xy plane is caused both by the random relaxation field and by the fact that the homogeneity of the magnetic field within the sample is not perfect, so that nuclei in different parts of the sample have different precession frequencies. There is an important difference between these two relaxation processes: the first is entirely random, and so unpredictable; whereas the second acts continuously and is

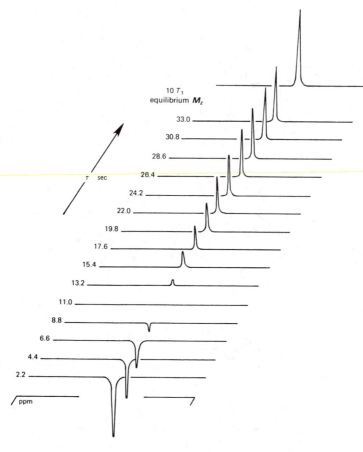

10 T_1
equilibrium M_z

33.0

30.8

28.6

26.4

τ sec

24.2

22.0

19.8

17.6

15.4

13.2

11.0

8.8

6.6

4.4

2.2

ppm

Fig. 5.25   The full population inversion experiment. A series of spectra are obtained at different τ. A plot of their intensities as a function of time gives the rate of relaxation, from which T_1 can be derived. (After Martin *et al.*, *Practical NMR Spectroscopy*. John Wiley and Sons Inc., New York, reprinted with permission.)

constant at each part of the sample, so that we can in principle correct for this relaxation contribution. We assume that the random contribution to relaxation is negligible. If we apply a 90° pulse, all the magnetization M_{xy} is the xy plane, but spins in different parts of the sample have different angular velocities due to the inhomogeneities in the magnetic field, and so some move ahead of the average and some lag behind. We wait a period of τ seconds for the spin distribution to evolve; the value of M_{xy} will decrease, and may even become zero if the field inhomogeneity is large. We then apply a 180° pulse, and all spins precess around B_1 to the other side of the xy plane. They do not take up their mirror positions. We have thus put the faster spins at the rear and the slower spins in

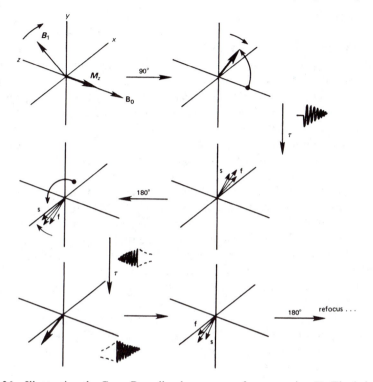

Fig. 5.26   Illustrating the Carr–Purcell pulse sequence for measuring T_2. The behaviour of the spins is shown relative to B_1, as if they were stationary. The 90° pulse produces M_{xy}, which then decreases as the spins move apart, s = slow, f = fast. The 180° pulse alters the position of the slower and faster spins, which now close up again and M_{xy} increases and then decreases.

front. They continue to precess, but after further time τ seconds they come into phase again and the magnetization M_{xy} is again a maximum. The 180° pulse is a refocusing pulse and the intensity of the spectrometer output rises following this pulse to a maximum τ seconds after and then decays again. Another 180° pulse will refocus the magnetization, which indeed can be refocused indefinitely, given perfect pulses. The refocusing does not work for the random relaxation process, and so the duration of the experiment is limited by the intrinsic transverse relaxation. Indeed, the echoes decay in intensity at a rate determined by the real T_2, which can be measured from a plot of intensity versus time. The experiment is summarized in Fig. 5.26. Each echo is effectively two back-to-back FIDs. It is possible to split them and Fourier transform each so that, if the sample contains several resonances, then T_2 can be measured separately for each from the resulting absorption spectra. If spin–spin coupling is present, however, then this does not work and modulation of the echoes is produced, which as we shall see proves useful in multidimensional spectroscopy.

Questions

5.1. In the inversion-recovery experiment used to measure T_1, the signal intensity is negative for short recovery times but positive for long recovery times. Calculate, in terms of T_1, at what recovery time the signal intensity is zero.

5.2. In Fig. 4.11, it is shown that for $B_0 = 8.48$ T and $10^3/T = 2.85$, the ^{195}Pt relaxation rate in $Pt(CN)_4^{2-}$ is given by $\ln R_1 = 0.8$. Calculate the ^{195}Pt linewidth of this anion under these conditions assuming $T_1 = T_2$ and a perfectly homogeneous magnetic field. The measured value of the linewidth is, in fact, 90 Hz. Calculate T_2 and estimate the contribution of scalar relaxation to this value.

5.3. It is required to run a series of spectral accumulations for a nucleus and a group of its compounds where the range of chemical shifts is likely to be 150 ppm . In order to accommodate all resonances without folding, a spectral width is chosen of 200 ppm. The spectrometer frequency is 90 MHz. What dwell time should be used? The longest relaxation time that is likely to be encountered is 20 s. A resolution of about 1 Hz is required. Calculate the appropriate memory size to be used to accumulate data (1K, 2K, 4K, 8K, 16K, 32K, 64K,...., where $1K = 1024$ are the permitted memory sizes) and the time required to transverse the memory and collect each FID following the read pulse. Will the nuclei be fully relaxed when the memory has been traversed and should the pulse length of the read pulse be 90° or less? What is the maximum permissible length of the stimulating read pulse if the nuclei are to be turned towards the xy plane by an equal amount whatever their chemical shifts?

6 NMR spectra of exchanging and reacting systems

6.1   Systems at equilibrium

One of the most important contributions that NMR has made to chemistry is the insight it has given into the dynamic, time-dependent nature of many systems, particularly those which are at equilibrium or where simply intramolecular motion is involved. Spectroscopy based on higher-frequency radiation, such as classical infrared (IR) or ultraviolet (UV) spectroscopy, has given mostly a static picture because the timescale of many processes is slow relative to the frequency used. However, the lower frequencies used for NMR and the smaller line separations involved, coupled with the small natural linewidths obtained, means that many time-dependent processes affect the spectra profoundly. As an example, we consider the spectroscopic behaviour of ethanol (EtOH) in Fig. 6.1. The proton spectrum of highly purified ethanol, $HO.CH_2CH_3$, is a methyl triplet due to coupling to the CH_2, an OH triplet for the same reason and a doublet of quartets for the methylene protons, which because of overlap approximates a quintet. Any acidic impurity catalyses interchange of OH protons between molecules:

$$EtOH + EtOH^* \xrightleftharpoons{H^+} EtOH^* + EtOH$$

The exchange of the protons results in a short break in the CH_2OH coupling path, and, since the total spin of the CH_2 protons in the two molecules between which the proton jumps may not be identical, then some of the OH protons will suffer an abrupt change in frequency. The result is to introduce uncertainty into their nuclear frequency and thus line broadening. This is the case in the spectrum of Fig. 6.1(a), which is of neat but not particularly well purified ethanol, in which the broadening of the OH and CH_2 proton signals is evident. Addition of acid causes acceleration of the rate of proton exchange to the extent that the OH–methylene coupling is completely lost and only an average frequency can be detected. The lines are now sharp. This is called the fast exchange region. Where the lines are broadened is called the region of intermediate exchange rates and where coupling is fully developed (not shown) is called the slow exchange region.

Ethanol also allows us to demonstrate another aspect of fast exchange. In the neat solution there is extensive hydrogen bonding between OH oxygen in

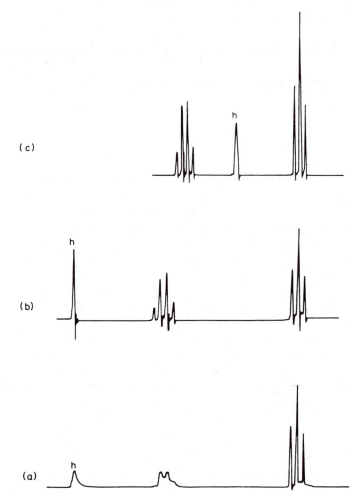

Fig. 6.1   The 60 MHz proton spectra of ethanol. (a) Neat, showing intermediate rate of exchange. (b) Neat, with a drop of HCl, causing fast exchange and loss of all OH–CH$_2$ coupling. (c) Diluted to 7% in CDCl$_3$, which produces an OH proton chemical shift. The OH resonance is labelled h.

one molecule and OH hydrogen in another. This interaction causes a low-field shift of the OH proton resonance. If the alcohol is diluted (Fig. 6.1(c)) then the hydrogen bonds become dissociated to an extent depending upon the dilution, and there is an up-field shift of the OH resonance. The solution now can be regarded as containing two types of ethanol, hydrogen-bonded and non-hydrogen-bonded. These two species have different OH proton chemical shifts but are not observed separately because of the fast OH proton exchange, which results in a signal of the average frequency being observed weighted by the concentrations of the two species. Thus chemical shift–dilution plots give information about the hydrogen bond dissociation.

Finally, we should note that in compounds such as ethanol there is very rapid rotation around the C–C bonds. Instantaneously, the molecule will have a particular configuration in which the C–O–H bond is bent and in which one of the CH_2 protons and one of the CH_3 protons will be nearest to the OH proton and so have a different chemical shift to its neighbours on the same carbon atom. These differences are not observed in the spectrum, however, because each proton has an average chemical shift due to the rapid internal rotation. In certain molecules such conformational changes can be quite slow and then the effects of the motion can be detected in the NMR spectra. Following the Karplus relationship (Fig. 3.1), the interproton coupling constants will also all be different in an ethyl group: these are also averaged by the rotation to the value of the average of the Karplus curve.

6.1.1 The effects of exchange on NMR spectra

We have described exchange by such words as 'fast' and 'slow'. It is now necessary to determine the timescale within which we can apply these terms correctly.

Consider that we have a sample that contains two types of spin a and b (of the same nucleus) with angular frequencies ω_a and ω_b. Initially, all the spin magnetization is disposed in the z direction. We will also suppose that there is no relaxation of the spins. If now we apply a 90° pulse, we will swing all the spins into the xy plane. If we place ourselves in a frame of reference rotating at the same angular velocity as spins a, these will appear to be stationary. Spins b, on the other hand, will rotate with an angular velocity of $\omega_b - \omega_a$, and after some time t the spins b will have moved an angle $\theta = (\omega_b - \omega_a)t$ away from spins a. In the diagram, we assume that $\omega_b > \omega_a$.

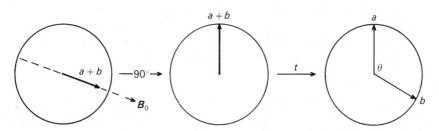

Now we will allow interchange of spins a and b and see what happens to the total magnetization of spins a. If there are N_a spins a and N_b spins b, and a small proportion E interchange in one second, then we can define an average lifetime (seconds) for an individual spin as

$$\tau_a = N_a/E \quad \text{and} \quad \tau_b = N_b/E$$

Two things happen at site a: spins are lost to site b and these spins are replaced by spins from site b which start immediately to precess at ω_a. However, because they were previously precessing at the different ω_b, they arrive in a position shifted relative to the spin a vector, and, as time proceeds, they form a fan in the xy plane.

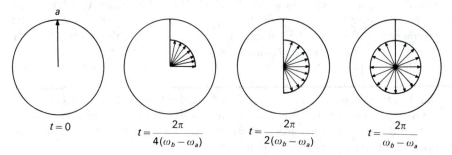

The arriving b spins thus take up all directions relative to the a spins and contribute nothing to the a magnetization, which thus decreases by what is effectively a T_2-type mechanism. If we assume that there is only a small change in magnetization per period $1/(\omega_b - \omega_a)$, then we can calculate a characteristic time for the decay of the a spin magnetization. Thus the number of b spins arriving at a in time t is Et, and the average value of the a magnetization is $\mu(N_a - Et)$, where μ is the magnetic moment of a nucleus. Replacing E by N_a/τ_a we have magnetization

$$\mu(N_a - N_a t/\tau_a)$$

The ratio between magnetization at time t and at time zero (μN_a) is

$$(1 - t/\tau_a)$$

Now the decay will not be linear with time since magnetization b is also decaying, so that as a decays the number of coherent b spins arriving at a also falls. While the rates of decay of a and b magnetization are not necessarily the same, the rate of loss of magnetization falls in proportion to its current value, and so can be regarded as exponential. The ratio of magnetization at time zero, M_0, and that at time t, M_t, thus follows the law

$$M_t/M_0 = e^{-t/T}$$

If t/T is small the exponential approximates to $1 - t/T$. Evidently, T is the characteristic time of the decay, and it follows that τ_a is its equivalent, or, in other words, can be regarded as equivalent to a relaxation time. So far, we have ignored the intrinsic relaxation of the spins. If we now introduce this with a characteristic time T_{2a}, then the overall apparent relaxation time T_{2a}^* is given by

$$1/T_{2a}^* = 1/T_{2a} + 1/\tau_a \tag{6.1}$$

A similar expression is obtained for spins b.

If N_a and N_b are equal, then the line broadening due to exchange will be the same for both lines. If the populations of the two types of spin are different, then the lifetimes will be different and the line broadening unequal. This type of behaviour is typical of slow exchange. It is defined by there being only a small proportion of the total nuclei exchanged in the time required for spins b to rotate one radian relative to spins a. The number of spins exchanging then is related to N_a by

$$E/(\omega_b - \omega_a) \ll N_a$$

which gives

$$\tau_a(\omega_b - \omega_a) \gg 1$$

The condition for fast exchange is, not surprisingly, the inverse of this relationship, i.e. the inequality sign is reversed

$$\tau_a(\omega_b - \omega_a) \ll 1$$

which implies that the spins all interchange position several times within one radian of displacement of ω_b from ω_a. In such a case, the individual behaviour of the two chemically shifted types of spin is lost and a singlet is obtained with properties that are the average of the two. Nor, provided the exchange is fast enough, does the exchange introduce any relaxation effects. If the exchange is somewhat slower, however, the fact that it is a random process does start to introduce T_2-type relaxation. To understand this, we have to consider the behaviour of a single spin. This will precess for short periods of time at either ω_a or ω_b and so will precess at the average of the two frequencies, weighted for the relative proportions of the two types of spin, P:

$$P_a = N_a/(N_a + N_b) \qquad P_b = N_b/(N_a + N_b)$$

The frequency of the singlet is then

$$\omega_{av} = P_a\omega_a + P_b\omega_b$$

or in chemical shift terms

$$\delta = P_a\delta_a + P_b\delta_b$$

If the relaxation times in the two different environments are different, then these are averaged in the same way.

If the exchange is slow enough that appreciable periods of time are spent precessing at either frequency during a time $1/(\omega_b - \omega_a)$, we can expect that each spin in the sample will behave differently, or detectably so, and so have a slightly different frequency, and that this will lead to dephasing and line broadening. We can see qualitatively what happens as follows. If $P_a = P_b$ and the lifetime between exchanges is τ, then there are on average $1/\tau(\omega_b - \omega_a)$ changes of frequency per radian rotation of b relative to a. Thus there are on average $1/2\tau(\omega_b - \omega_a)$ short periods spent at each of the frequencies ω_a and ω_b. Because these periods are of random length, then we can assume that a number θ at ω_b is missed and the extra number are spent at ω_a. This will give some idea of the likely spread of frequencies. The average frequency is then

$$\omega_{av} = \frac{[1/2\tau(\omega_b - \omega_a) + \theta]\omega_a + [1/2\tau(\omega_b - \omega_a) - \theta]\omega_b}{1/\tau(\omega_b - \omega_a)}$$

which on rearranging gives

$$\omega_{av} = (\omega_a + \omega_b)/2 + \tau\theta(\omega_b - \omega_a)(\omega_a - \omega_b)$$

A spread of frequencies around the average $\frac{1}{2}(\omega_a + \omega_b)$ is thus introduced by the random nature of the exchange process, which is proportional to $\tau(\omega_b - \omega_a)^2$.

The full expression for non-equal populations, which is obtained by deriving the behaviour of the magnetization under exchange, is

$$1/T_2^* = P_a/T_{2a} + P_b/T_{2b} + P_a^2 P_b^2 (\omega_b - \omega_a)^2 (\tau_a + \tau_b) \tag{6.2}$$

We thus see that, in the slow exchange region, the two resonances are detected, and that these broaden as the rate of exchange increases. In the fast exchange region, the two resonances give a single, narrow coalesced signal, which broadens as the rate of exchange decreases. The question now is what happens at intermediate rates of exchange where

$$\tau(\omega_b - \omega_a) \approx 1$$

On the fast exchange side of this condition, we have a broadened singlet, and on the other side, we have a broadened doublet, so that a quite dramatic change in spectral shape must take place as the exchange rate passes through this condition. Reference to equations (6.1) or (6.2) will show that the linewidth becomes comparable with the frequency separation of the signals in the slow exchange limit, so evidently we will see a maximum width of the resonance. The relaxation in fact becomes non-exponential, so that we cannot use either of the above simple approaches to try to work out what may happen. We can nevertheless use the construction below in order to obtain some information.

As before, we turn all the magnetization into the xy plane by using a 90° pulse. After the b spins have rotated relative to the a spins by about one radian, most of the original a spins will have interchanged with b spins and the a magnetization will almost have disappeared. The arriving b spins will form a fan, but this will diminish in magnitude very rapidly since the b spins are also

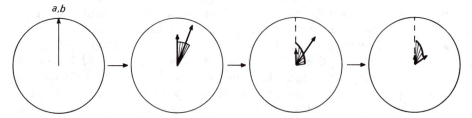

losing intensity at a similar rate to the *a* spins. The magnetization of the arriving *b* spins thus will never average to zero and will always have a resultant that will be displaced from the *a* magnetization towards the position of the *b* magnetization. The total magnetization will then move increasingly away from the position of the original *a* magnetization, thus pulling the vector to move faster towards the *b* spins and so to increase in frequency. Similarly, the frequency of the *b* spins apparently decreases. In addition, there is an increasingly rapid decay of the magnetization, which gives the signal a rather square-topped appearance (see Fig. 5.7(b)). The frequency (or chemical shift) separation of the two signals is thus reduced, so that the lines broaden as the rate of exchange is increased from the slow exchange region, start to move together and eventually coalesce into a broad, flat-topped singlet, which then narrows with further increase in exchange rate. Typical behaviour of this sort is illustrated in Fig. 6.2, where the rate of rotation of the *N,N*-dimethyl group of dimethylformamide is monitored as a function of temperature. In the quasi-static situation at 35° the two methyl groups have different chemical shifts and two signals are observed. At higher temperatures, the barriers to rotation are much less effective, and coalescence to a singlet occurs. In principle, if a computer fit is made to the line shapes, it is possible (using the full theory) to extract values of τ at all temperatures. Unfortunately, the spectral changes are significant over a rather small temperature range of some 26°C, so that the method is of limited accuracy if it is wished to obtain activation parameters. The accuracy can, however, be improved in a variety of ways.

We should first note how accurately defined is the coalescence point at 118°C. It can be shown that, at coalescence, we have the relation

$$\tau_{co} = 2\sqrt{2}(\omega_b - \omega_a)$$

where

$$2/\tau_{co} = 1/\tau_a + 1/\tau_b$$

We can thus relate τ to temperature very accurately at coalescence and can, for example, obtain a value for the free energy of activation at that temperature using the Eyring theory

$$\Delta G_T^* = 8.31\,T[23.76 + \ln(T/k)]$$

where the rate constant k is obtained from $k = 1/\tau_{co}$.

Alternatively, we can obtain data over a wide range of temperature if we can monitor coalescence points for a system at different spectrometer frequencies (or magnetic field strengths) and for different nuclei or groups with different chemical shifts that are affected equally by the exchange. This approach has been used to study hindered rotation of the *N*-ethyl groups of the iron complex $(CH_3CH_2)_2N-C(S)SFe(CO)_2(C_5H_5)$. Both the methyl and methylene protons of the ethyl groups are non-equivalent at low temperatures, but the chemical shift between methyl signals is less than that between methylene signals, so that

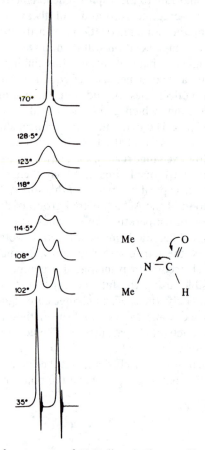

Fig. 6.2   The *N*-methyl resonance of *N,N*-dimethylformamide neat liquid at different temperatures. Rotation of the Me$_2$N groups around the CN bond is slow at room temperature due to partial double-bond character in the bond, (see inset). At high temperatures the barrier to rotation is overcome and the two methyl groups see the same average environment. The spectrum at 118°C, where the doublet structure is just lost, is said to be at coalescence. (Reproduced with permission from Bovey (1965) *Chem. Eng. News*, 30 Aug., 103; copyright (1965) American Chemical Society.)

we can monitor two coalescence points. In addition, we can obtain two more coalescence points from the ^{13}C spectra of the two groups, and by using two different spectrometers we can observe a total of eight coalescences. The results are shown in Fig. 6.3, which gives the Arrhenius parameters $E_a = 66\,\mathrm{KJ\,mol^{-1}}$ and $\log A = 13.1$.

A more complex example of the hindered rotation in amides is shown in Fig. 6.4 for the *cis* and *trans* isomers of a vinyl diamide. Only the methylene

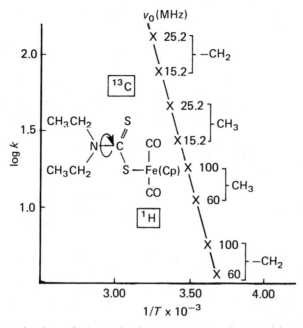

Fig. 6.3 Determination of the activation parameters characterizing the hindered rotation around the C–N bond of $(CH_3CH_2)_2$ N–CS–S–Fe(CO)$_2$Cp (Cp = cyclopentadienyl). (From Martin *et al.* (1980) *Practical NMR Spectroscopy*; copyright (1980) John Wiley and Sons Inc., New York, reprinted with permission.)

proton resonances are shown in the figure, and they indicate that several different rotation processes take place. If we take the *trans* isomer first, we see that at the highest temperature recorded there is a well resolved methylene quartet overlying a broadened resonance. This latter broadens and splits at lower temperatures with a coalescence temperature of 345 K. Thus one of the amides is rotating more slowly than the other, which only shows coalescence at 295 K. Exchange is no longer evident in the spectra at 228 K. Only three quartets are observed because of overlap of the two low-field quartets. In order to understand the spectra completely, it is necessary to identify which amide resonance is which, and this is done using nuclear Overhauser and two-dimensional proton–carbon correlation spectroscopy, which we will describe later. It is sufficient to record here that it is the amide on the CMe carbon that is rotating faster. The *cis* isomer shows much more complex behaviour, though at the higher temperatures there are two corresponding coalescence points at 355 K and 333 K. The quartets, however, split further at lower temperatures, and each resonance is transformed into a doublet of quartets due to the introduction of extra spin–spin coupling between geminal protons on the same carbon atom. The methylene protons have thus been rendered non-equivalent by some further restriction in

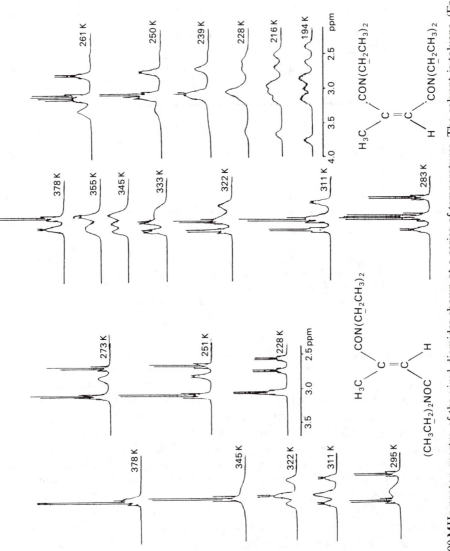

Fig. 6.4 The 300 MHz proton spectra of the vinyl diamides shown at a series of temperatures. The solvent is toluene. (From Szalontai et al. (1989) *Magn. Reson. Chem.*, **27**, 216–22; copyright (1989) John Wiley and Sons Ltd, reprinted with permission.)

the motion of the molecule. The activation parameters were obtained by calculating τ at the coalescence points. To do this accurately, it is of course necessary to know the chemical shift between the two coalescing signals. This may vary with temperature, and for the present example it was found necessary to measure the chemical shifts in the slow exchange limit over a wide range of temperatures to ensure that the correct frequency separation had been used.

The other factor that weighs heavily on the accuracy is of course the temperature of the sample. This is controlled by constructing a Dewar shield around the sample holder and passing nitrogen gas over the sample at a known temperature. This is regulated by a thermocouple placed near the sample tube. The method is prone to temperature gradients and so inaccuracies, which are minimized by calibrating the temperatures using known standards, for instance the chemical shift between the OH and CH_3 resonances of a carefully prepared methanol sample.

One very wide field of interest is that of ligand exchange on complex cations. These processes have been studied in a variety of ways, and the rates measured for ligand exchange vary over many orders of magnitude and depend upon the nature of the cation chosen for study. In many instances, the rates fall, or can be made to fall by changing the temperature, in the range accessible to NMR studies, so that this relatively new technique has been used to extend our knowledge of such systems. In addition, since NMR provides the means to study systems at equilibrium, our studies can be of the unperturbed system.

A particularly well studied system is that of solvent exchange on the cation $Al(H_2O)_6^{3+}$ in aqueous solution. This cation contains the NMR-active nuclei ^1H, ^{27}Al and, if isotopically enriched, ^{17}O, all of which have been used in its study. Such a highly charged ion is subject to hydrolysis and so solvent exchange can be quite a complex process:

$$Al(H_2O)_6^{3+} + H_2O^* \xrightleftharpoons{k_{ex}} Al(H_2O)_5H_2O^* + H_2O$$

$$H^+\uparrow\downarrow + H^+ \qquad\qquad\qquad -H^+\uparrow\downarrow + H^+$$

$$Al(H_2O)_5(OH)^{2+} + H_2O^* \rightleftharpoons Al(H_2O)_4(H_2O^*)(OH)^{2+} + H_2O$$

The hydrolysis reaction causes exchange of hydrogen atoms, which in the present case is very fast, although the backward reaction is much faster than the forward reaction and the hydrolysis constant is only 10^{-5} ($pK = 5$). The oxygen exchange is much slower, though two pathways are possible via either the hydrolysed or the non-hydrolysed ions. We would like to know the rates of exchange of both hydrogen and oxygen, whether the latter is influenced by the acidity of the solution and so whether the hydrolysed species contributes to the exchange, and, if possible, the mechanism by which a water molecule replaces a complexed water molecule.

The rate of proton exchange can be studied using ^1H spectroscopy. If a solution of an aluminium salt is cooled, this slows down the rates of exchange.

Fig. 6.5   The ^{1}H spectrum of 3 M AlCl$_3$ at $-47°$C. The water complexed by the cation is seen 4.2 ppm to low field of free, bulk water and is indicated by the letter s. The spectrum illustrated was obtained at 60 MHz. (After Schuster and Fratiello (1976) *J. Chem. Phys.*, **47**, 1554, with permission.)

If the temperature is low enough, and this can be achieved either by adding an antifreeze such as acetone or by using a very concentrated salt solution, then two proton resonances are observed due to solvent and complexed water, as shown in Fig. 6.5. Integration of the resonances together with knowledge of the aluminium concentration allows the hydration number (the number of water ligands attached to the cation) to be determined, and this is approximately equal to six, the method being less precise than the isotope fine-structure method depicted in Fig. 2.8. The chemical shift between the two resonances is 4.2 ppm, and we can calculate that coalescence will occur when the lifetime of the protons is about 1.8×10^{-3} s. We can therefore study the rate of exchange in dilute solution by measuring the proton relaxation times as a function of temperature and acidity, and this gives a protolysis rate constant of 0.79×10^5 s^{-1} at 298 K. The ^{27}Al resonance is also influenced by this process since the hydrolysed cation has a much broader resonance than that of the non-hydrolysed species. The latter has a width of some 2 Hz, but the fast exchange mixes in the rapid quadrupolar relaxation of the hydrolysed cation and gives lines that are generally in the range 10–20 Hz wide. Note that, in this case, the exchange of one atom (H$^+$) results in apparent exchange of a second (^{27}Al).

The rate of oxygen exchange is best studied using the ^{17}O resonance. Here, the rate of exchange was known to be slower, so that resonances should be observed for bound and bulk water molecules. In fact, the chemical shift between the two is very small and, because of the quadrupolar broadening of the lines, they are not resolved. Separating the two resonances has been done in a variety of ways, but the one used for the most comprehensive studies was to add a paramagnetic cation, Mn^{2+}, for which the whole water molecule exchange rate is very fast. The water not attached to aluminium thus is subjected rapidly to the very large magnetic field of the electron spins on the Mn^{2+}, and has its T_2 very much reduced and so its linewidth increased, to the extent that its signal disappears in the baseline of the spectrum. Only the oxygen bound to Al^{3+} is observed and, provided the exchange on Al^{3+} is slow enough, its relaxation is determined entirely by its own natural relaxation time and exchange lifetime (Equation 6.1)). The problem in making such measurements with quadrupolar nuclei is that the relaxation time in the absence of exchange is not known and that it is very temperature-dependent. The relaxation time, or linewidth, thus has to be measured over a range of temperature wide enough that some data are acquired in the region where the exchange broadening is negligible. A set of measurements is shown in Fig. 6.6, where it will be seen that the temperature dependence of relaxation falls into two regions: lower temperatures, where the linewidth is dominated by the quadrupolar mechanism, and higher temperatures, where it is dominated by exchange. Extrapolation of the quadrupolar influence in principle allows the exchange rates to be determined as a function of temperature. In fact, because of the curvature, it is difficult to estimate the slope of the line giving k_{ex} with the required precision, and a separate experiment was needed in order to increase the range of values of k_{ex} observed. This was done using an injection or stopped-flow technique in which an aluminium salt solution in H_2O was injected rapidly into acidified H_2O enriched in ^{17}O, both solutions being at low temperature around 256 K. Mn^{2+} was also present in the ^{17}O water. Thus, as the water exchanged with the aluminium solvation sphere, the ^{17}O signal of the mixture increased in intensity and the exchange rate could be obtained by plotting this as a function of time. This isotope exchange type of experiment is a common way of investigating systems that are at equilibrium. One is effectively perturbing the system but the perturbation is slight. The combined linewidth and stopped-flow results are shown in Fig. 6.6(b). In the case of aluminium it is possible to suppress hydrolysis completely by the addition of sufficient acid. Thus the data gave thermodynamic parameters for oxygen exchange on the non-hydrolysed cation: $k = 1.29 \, s^{-1}$, $\Delta H^* = 84.7 \, kJ \, mol^{-1}$, $\Delta S^* = +41.6 \, J \, K^{-1} \, mol^{-1}$. In addition, by measuring the exchange rates as a function of pressure at constant temperature, it was possible to obtain an activation volume for the oxygen exchange: $\Delta V^* = +5.7 \, cm^3 \, mol^{-1}$. We will discuss such experiments in a little more detail later, but the inference is that the transition state during the exchange involves a short-term increase in total volume. This occurs if the exchanging ligand leaves before the entering solvent molecule takes its place; in other words, it is a dissociative mechanism of exchange. In fact, the ΔV^* term is a little smaller than theory would demand if there were full dissociation, and the mechanism probably involves interchange during the dissociation, commonly called the I_d mechanism.

In the case of the gallium cation, $Ga(H_2O)_6^{3+}$, the hydrolysis of the ion is too strong to be suppressed completely by the addition of acid, and in this case oxygen exchange on both hydrolysed and non-hydrolysed species has to be measured. It is found that

$$k_{ex} = k_1 + k_2/[H^+]$$

where $k_1 = 403 \, s^{-1}$ and $k_2 = 14 \, mol \, s^{-1}$, both at 298 K.

If the rate of exchange is slow, so that linewidths are not significantly

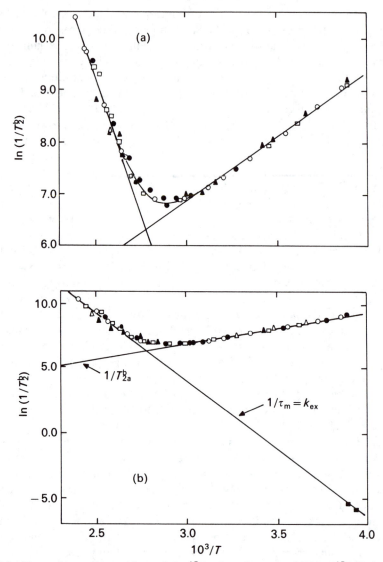

Fig. 6.6   Plots of natural logarithms of the ^{17}O relaxation rates $(1/T_2)$ of $^{17}OH_2$ bound to the aluminium cation, as a function of the reciprocal temperature. The T_2 values were obtained from the ^{17}O linewidths. (a) How the linewidth decreases as the temperature is increased (from the right-hand side of the plot) and how these data can be extrapolated satisfactorily into the exchange perturbed region. The onset of detectable exchange causes the plot to curve upwards as the temperature is increased beyond 333 K. (b) This contains the same data but includes some fast injection data also. (From Merbach *et al.* (1985) *Helv. Chim. Acta*, **68**, 545, with permission.)

perturbed, but where an injection-type experiment is not feasible, such as for an intromolecular rearrangement, then other means of studying the system have to be sought. The trick in this case is to perturb the nuclear spins in some way and observe how the spin populations recover under the influence of exchange as well as of relaxation. There are several versions of this type of experiment, but two examples will suffice to illustrate the possibilities.

The first is selective spin inversion followed by magnetization transfer. The iron dihydro complex of the ligand $P(CH_2CH_2CH_2P(CH_3)_2)_3$ has the structure depicted in Fig. 6.7. This structure can be established on the basis of the ^1H and ^{31}P spectra. There are two hydride signals at -12.8 and -14.1 ppm (note the very high-field shifts), which are split by spin coupling to each other and to all four phosphorus atoms into a 24-line multiplet consisting of a doublet of doublets of doublets of triplets, proving that two ^{31}P nuclei are equivalent. There are three ^{31}P signals in the intensity ratio 1:2:1, but these are broadened by an exchange process, which also causes slight broadening of the hydride proton resonances. If the influence of the ^{31}P nuclei is removed from the ^1H spectrum by double irradiation (see Chapter 7), the hydride resonances become a pair of doublets, $^2J(HH) = 13.3$ Hz, which are well separated and so amenable to study. We measure their rate of exchange by manipulating the spins using a pulse sequence and seeing how the resulting spectra vary as a function of a time interval. First, we apply a $90°$ B_1 pulse, which swings the magnetization of both types of proton into the xy plane. If we label the protons A and B, then the magnetizations are M_z^A and M_z^B. We then wait a time $1/2(v_A - v_B)$, where v_A and v_B are the frequencies of the two types of hydride proton at the spectrometer frequency used. During this time M^A and M^B precess at different

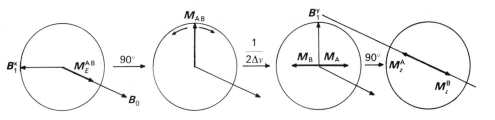

frequencies in the xy plane and at its end are displaced from each other by $180°$; in other words, they are in line. The relatively small coupling interaction has a negligible effect. Next we change the phase of the B_1 signal by $90°$, which has the effect of moving the B_1 vector $90°$ in the xy plane in the rotating frame. A $90°$ pulse then swings the magnetization back into the z direction, with the difference that the magnetization of one type of hydride is pointing in the opposite direction to its equilibrium direction and has been inverted. We then apply a further $90°$ read pulse after some time interval t and produce a FID. The response of the spins that have been returned to their equilibrium direction

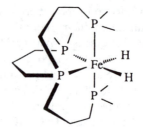

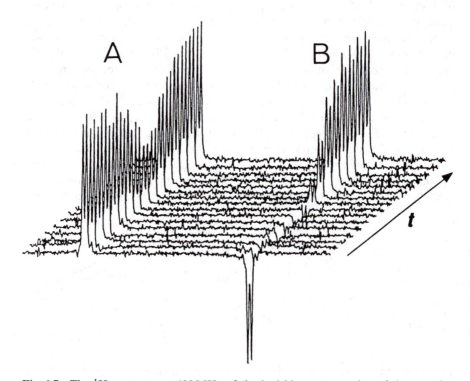

Fig. 6.7   The ^{1}H spectrum at 400 MHz of the hydride proton region of the complex depicted. The magnetization of one of the types of hydride proton was inverted selectively using the pulse sequence discussed and the way the spectrum varied with time t after inversion was observed. The effect of the ^{31}P spins was removed by double irradiation. (From Field *et al.* (1991) *Magn. Reson. Chem.*, **29**, 36; copyright (1991) John Wiley and Sons Ltd, reprinted with permission.)

will, in the absence of exchange, give a signal after Fourier transformation that does not alter with t. The inverted spins will behave as in the inversion-recovery experiment illustrated in Fig. 4.1, and we can thus measure T_1 for these spins, observing them decay from a negative signal through zero to a positive signal. In the presence of exchange, however, the two types of behaviour are mixed and the time dependence of the two responses depends upon both T_1 and k, the rate constant for the exchange. The rate of change of the magnetization of the two types of spin can be expressed as two differential equations:

$$\frac{dM_z^A}{dt} = \frac{-(M_z^A - M_0^A)}{T_1} - kM_z^A + kM_z^B$$

$$\frac{dM_z^B}{dt} = \frac{-(M_z^B - M_0^B)}{T_1} - kM_z^B + kM_z^A$$

where the M_0 are the equilibrium values of magnetization and T_1 is the relaxation time, which is assumed to be equal for both types of spin. The first part of these two equations represents simply the normal exponential T_1 relaxation rate of the magnetization discussed previously, and the kM_z terms mix in the exchange process. Since M_z^A and M_z^B are initially of opposite sign, the exchange causes significant changes in the spectra, as shown in Fig. 6.7. As labelled, the B spins have been inverted and the A spins are normal. The replacement of A spins by B spins reduces the A magnetization and the A signal decreases in intensity. However, as the B spins approach their normal state, their effect becomes smaller and the A signal resumes its normal intensity. The B signal is initially inverted but its intensity is very rapidly reduced towards zero both by relaxation and by their replacement of B spins by A spins. The recovery to normal is thus more rapid than if controlled solely by relaxation. By fitting the evolution of the signal intensities with time to the differential equations, the two unknown quantities T_1 and k can be determined. Values obtained at 275 K are $k = 2.92\,s^{-1}$ and $T_1 = 1.73\,s$. The mechanism of the exchange is not known. It is likely to proceed via a five-coordinate intermediate in which either an Fe–H or an Fe–P bond is broken. The fact that the H–H coupling is maintained does, however, suggest that the Fe–H bonds remain intact and thus that the rupture of a bond to the terminal phosphorus atoms permits the hydride hydrogen atoms to change positions relative to the tertiary phosphorus atom and so change chemical shift.

It is equally possible to carry out this experiment by using a long selective pulse to invert one of the types of spin and then monitoring the spectra as a function of time. The choice of method depends upon what facilities are available with a given instrument.

A second example concerns the rate of slow ring flexing in *cis*-decalin, shown in Fig. 6.8. The motion causes the exchange of chemical shift of carbon atoms 2 and 3 and of 6 and 7. The lower sensitivity of ^{13}C spectroscopy means that a simpler approach may be necessary that requires fewer spectra to be produced.

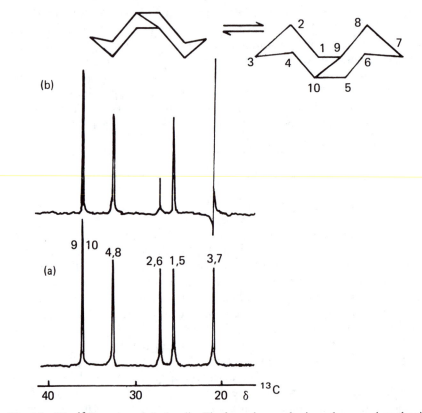

Fig. 6.8   The ^{13}C spectra of *cis*-decalin. The formula, numbering scheme and mechanism of exchange of the 2,6 and 3,7 carbon atoms are shown at the top of the figure. The spectra are obtained (a) normally and (b) with pre-irradiation of the 3,7 signal. The coupling effect of the protons in the molecule has been removed by broad-band double irradiation. (From Mann (1976) *J. Magn. Reson.*, **21**, 17, with permission.)

The ^{13}C spectrum of the compound is shown in Fig. 6.8(a) with the lines assigned to the appropriate carbon atoms. Two measurements are then necesssary: a determination of T_1 by the usual inversion-recovery experiment and then a selective irradiation experiment. A long selective pulse is applied to one of the resonances of the exchanging pair for a period of about T_1 seconds. This strongly perturbs the magnetization, so that M_z is reduced. Once equilibrium has been attained, the selective pulse is switched off and the spectrum obtained immediately by applying the usual 90° read pulse. The result is shown in Fig. 6.8(b), where the irradiated 3,7 signal is much distorted and the exchange-coupled 2,6 signal is much reduced in intensity. The intensities without irradiation, M_z^0, and with it, M_z^{irr}, are measured and the rate constant for ring flexing is then

$$1/k = [M_z^{irr}/(M_z^0 - M_z^{irr})]T_1$$

In the case of systems exhibiting fast exchange, it may simply be necessary to reduce the temperature sufficiently to slow down the exchange process and cause the slow exchange limit spectrum to appear, since this will be much more informative than the fast exchange spectrum, which may be just a singlet. A typical example is the ^{17}O spectrum of the cobalt carbonyl compound, $Co_4(CO)_{12}$. The carbonyl groups move around the cobalt cluster so that they all experience the full range of chemical environments and shifts in the molecule. Spectra at several temperatures are shown in Fig. 6.9, and it will be evident that the ambient-temperature trace is not very informative but that at $-25°C$ all four types of carbonyl group can be distinguished. Note that the ^{59}Co spectrum of this compound contains two resonances in the intensity ratio 3:1 so that the carbonyl exchange takes place on a cluster in which the positions and bonding of the cobalt atoms are invariant. The nucleus ^{59}Co is quadrupolar and the asymmetry of the environment around the metal atoms means that the relaxation time is very short. Linewidths are of the order of 7500 Hz but ^{59}Co chemical shifts are very large and so the resonances are resolved.

Another example is that of the behaviour of t-butyllithium, which demonstrates also the averaging effects of exchange upon coupling constants and coupling multiplicities. The compound was studied using ^{13}C spectroscopy and the spin-spin coupling of this nucleus to the lithium nuclei. The common isotope of lithium is ^{7}Li, which has a quadrupole moment sufficiently large that its quadrupole relaxation times are quite short, and so any coupling is not well resolved. On the other hand, the less abundant isotope ^{6}Li has a quadrupole moment that is some 50 times smaller, which is indeed the smallest among all the elements, and which thus behaves much more like a spin-1/2 nucleus than like a quadrupolar one. Further, ^{6}Li is easily available via the nuclear industry and so is proving to be a useful tool for the study of organolithium compounds by NMR methods. The structure of t-butyllithium is shown in Fig. 6.10, and the ^{13}C spectrum of the α-carbon atoms of a sample highly enriched in ^{6}Li is shown in Fig. 6.11 as a function of temperature. This compound forms tetramers in cyclopentadiene solution in which the lithium atoms are arranged in a tetrahedral cluster with the alkyl groups bonded to three lithium atoms by a multicentre bond and so situated over each face of the tetrahedron. The tetramer is, however, fluxional and the expected spectrum is only observed at low temperatures. The α-carbon atom of the t-butyl group is coupled equally to its three nearest-neighbour ^{6}Li atoms with a coupling constant of 5.4 Hz. For ^{6}Li $I = 1$ so that a seven-line multiplet results, with intensities in the ratio 1:3:6:7:6:3:1, and this is the spectrum observed at $-88°C$ in Fig. 6.11. The coupling to the distant lithium nucleus is undetectably small. As the temperature is increased, fluxion of the structure occurs, and at 26°C this is fast enough for the α-carbon atoms to be influenced equally by all four ^{6}Li nuclei, which produces a nine-line multiplet with relative intensities of 1:4:10:16:19:16:10:4:1. In addition, the close and long-distance coupling constants are mixed and the coupling constant has three-quarters of its low-temperature value at 4.1 Hz. Note

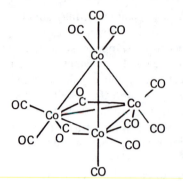

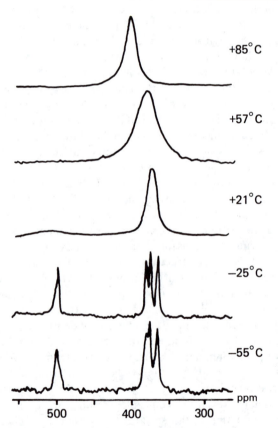

Fig. 6.9   The ^{17}O spectrum of $(CO)_{12}Co_4$. The carbonyl groups exchange positions, but this can be slowed down sufficiently by cooling to enable the four types of carbonyl group to be seen. The reference is water. (From Aime *et al.* (1981) *J. Am. Chem. Soc.,* **103**, 5920; copyright (1981) American Chemical Society, reprinted with permission.)

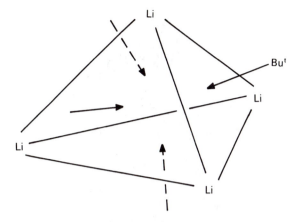

Fig. 6.10   The structure of tetrametic t-butyllithium. The lithium atoms are arranged in a tetrahedron with the alkyl groups bonded equally to three lithium atoms and so positioned above a face of the tetrahedron. Only one alkyl group is shown.

that there can be no intermolecular exchange since this would involve breaking the coupling path and would collapse the multiplet to a singlet.

A final example of fast exchange where the structure can be deduced only at low temperature is that of the intramolecular transfer of the $CuPEt_3$ group around the ring of the σ-cyclopentadienyl complex, σ-$C_5H_5CuPEt_3$. This compound, whose structure is shown in Fig. 6.12, contains three distinguishable types of hydrogen atom, and its proton spectrum should consist of three chemically shifted lines with intensity ratios 1:2:2. In fact, only a singlet is observed, even at $-1°C$, as shown in Fig. 6.13. Cooling causes broadening, until, by $-54°C$, the expected structure starts to emerge as the exchange is slowed down. Interestingly, there is an obvious asymmetry in the spectrum of the 2:2 doublet between -46 and $-56°C$, and this allows us to deduce the mechanism of the transfer of the $CuPEt_3$ group. There are three possibilities for exchange in complexes of this type: (i) 1:2 hops, where the $CuPEt_3$ group moves to the carbon atom to either side and its immediate neighbour: (ii) 1:3 hops, where it jumps to the next nearest neighbour on either side; and (iii) a random shift process, where both types of shift occur with equal probability. This is shown in Fig. 6.12(a), where the chemically shifted protons are lettered A, B and X. The table in Fig. 6.12(b) shows how the spin chemical shifts change for both the 1:2 and 1:3 processes. Thus, for the latter, one B becomes A, and the other remains B; one A becomes B but the other becomes X, and X in turn becomes A. In the case of the 1:3 hops, then, only A interchanges with X, whereas for the 1:2 hops it is B that interchanges with X. Since the chemical shift between A and B is small and that between either A or B and X is much larger, it is the interchange with X that causes most line broadening. We can see immediately from Fig. 6.13 that one of the A or B lines is more broadened

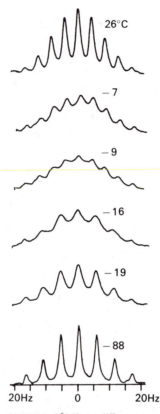

Fig. 6.11   The ^{13}C spectrum of $[(CH_3)_3C^6Li]_4$ at different temperatures in cyclopentadiene solvent. Only the resonances of the α-carbon atom are shown. The long-distance coupling to the methyl proton nuclei was eliminated by double irradiation at the proton frequency. (From Thomas *et al.* (1986) *Organometallics*, **5**, 1851; copyright (1981) by the American Chemical Society, reprinted with permission.)

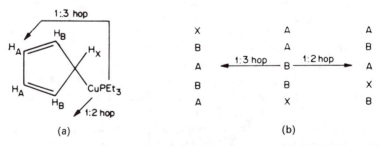

Fig. 6.12   The structure of the σ-cyclopentadienyl complex σ-$C_5H_5CuPEt_3$, with the protons of different chemical shifts labelled A,B,X. The paths of the different hopping processes are indicated and the table shows how each path causes different interchanges of proton position to occur. (From Whitesides and Fleming (1967) *J. Am. Chem. Soc.*, **89**, 2855; copyright (1967) American Chemical Society, reprinted with permission.)

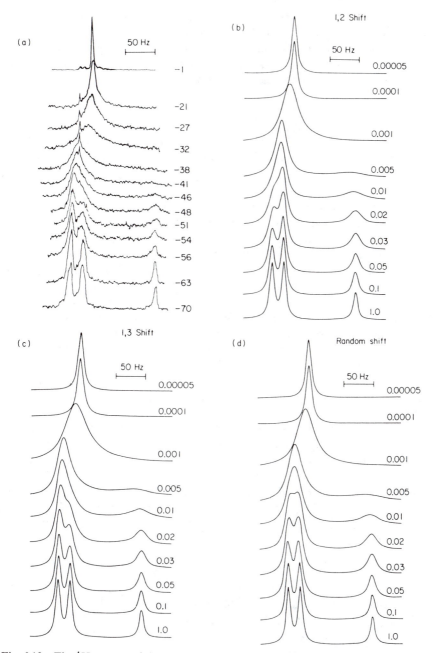

Fig. 6.13 The ^1H spectra of the complex σ-$C_5H_5CuPEt_3$ at temperatures between -1 and $-70°C$ together with spectra calculated for three possible mechanisms for $CuPEt_3$ migration over a range of residence times. (From Whitesides and Fleming (1967) *J. Am. Chem. Soc.*, **89**, 2855; copyright (1967) American Chemical Society, reprinted with permission.)

than the other, so that the process is not random. It is then necessary to determine which resonance is A or B, and this may be done by cooling sufficiently to find which of the two resonances is coupled most strongly to X. It is not always possible to achieve this, but where it has been done it is shown that it is the 1:3 mechanism that is the correct one.

6.2   High-pressure spectroscopy

While it is normal to study the dynamics of a system as a function of temperature in order to obtain the thermodynamic properties of the changes observed, it is equally valid to use pressure as the variable, as this may give further insight into the mechanisms of the reactions observed. Often, if a full study is required, both temperature and pressure have to be varied.

In order to work at high pressure, a special probe has to be constructed. There are several variations on the structures of such probes, depending upon exactly what type of experiment it is required to undertake. A typical assembly consists of a pressure containment vessel of a size such that it can be placed in the probe space of an NMR spectrometer and with electrical leads through to the sample coil and with an inlet for the pressure fluid. This vessel contains a sample space surrounded by a detector coil and a temperature measuring device such as a Pt resistance thermometer. The sample is enclosed in a small glass tube with a close-fitting plastic piston to transmit pressure to the sample. Resolution of such probes can be good, even without sample spinning, and temperature control is excellent since the heat transfer medium is the pressurizing fluid.

Not surprisingly, the effect of pressure on the hindered rotation of amides has been quite deeply studied. ^{1}H spectra of the compound N,N-dimethyltri-chloroacetamide are shown in Fig. 6.14 as a function of pressure at constant temperature. The expected doublet is coalesced near ambient pressure but splits as the pressure is increased. Thus the pressure decreases the rate of rotation, though there is relatively little change above pressures of 200 MPa. Now, the rate of hindered rotation in such compounds can be expressed in the usual transition-state form

$$k \propto e^{-E_0/RT}$$

where E_0 is some energy barrier. The pressure then has no direct effect on the amide rotation but operates via a change in the viscosity of the solvent, which increases with pressure. We thus have to write

$$k = F(\eta)e^{-E_0/RT}$$

where η is the shear viscosity and $F(\eta)$ is a function of η that does not vary linearly with either pressure or viscosity. It is possible to calculate $F(\eta)$ by making a variety of assumptions, and the results suggest that there is some

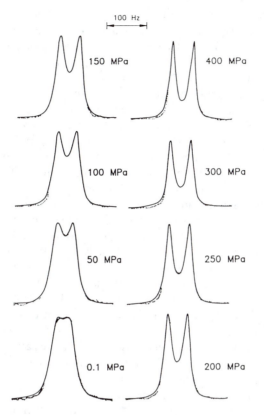

Fig. 6.14   The ^{1}H spectra of the N-methyl resonances of Me$_2$NC(O)CCl$_3$ dissolved in pentane at 282.3 K as a function of applied pressure. (Jonas *et al.* (1990) *J. Chem. Phys.*, **92**, 3736; reprinted with permission.)

form of coupling between the solvent motion and the rotation of the N,N-dimethyl groups.

If the exchange is very slow, so that the resonances of the various species present are well resolved, it may not be possible to obtain exchange rates, but the intensity of the resonances may vary, and this indicates a displacement of the equilibrium between the components, which can be caused by changes in either temperature or pressure. In the latter case, change occurs if there is a difference in volume between the species in equilibrium. As an example of such a study, we take the equilibrium

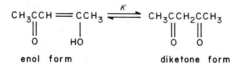

The value of the equilibrium constant K increases with pressure, so that the β-diketone is favoured at high pressures and its volume thus must be smaller than that of the enol. The difference in partial molar volumes is found to be $-4.5 \, cm^3 \, mol^{-1}$ and is believed to arise because of the extra volume occupied by the hydrogen-bonded ring formed by the enol.

The exchange of ligands on cations is also studied by high-pressure NMR. Here we will consider the two complexes of beryllium, $Be(DMSO)_4^{2+}$ and $Be(TMU)_4^{2+}$, where DMSO is dimethylsulphoxide, Me_2SO, and TMU is tetramethylurea, $(Me_2N)_2CO$. Both ligands bond via the oxygen. Exchange is studied in the presence of excess ligand (cf. the Al^{3+}–water system described above), but with organic ligands it is possible to dilute the system with an inert solvent such as deuteriated acetonitrile, CD_3CN, and vary the ratio between the concentrations of free ligand and complex and so obtain data about the order of reaction. By 'inert' we mean here a solvent that will not compete with the ligand for sites on the cation but which may nevertheless form complexes in an even more inert solvent. Such systems are most easily studied using ^1H spectroscopy of the methyl groups. Two signals are observed due to bound and free ligand, and the exchange rates can be found from the shapes of the spectra, which for much of this work demands the calculation of full spectral envelopes. In the case of the TMU complex, it was found that the rate of exchange was independent of the concentration of free ligand, so that the reaction is first-order and this implies that the ligand must dissociate from the cation before exchange can take place. In the case of the DMSO complex, the exchange rate did depend upon the concentration of ligand, so that the reaction is second-order, which implies that a ligand must associate with the $Be(DMSO)_4^{2+}$ before another ligand will leave. Variable-temperature determinations then permit the thermodynamic parameters ΔH^* and ΔS^* to be obtained as well as the rate constants, and these are given in Table 6.1. The different signs of the entropy values are also in accord with the different reaction mechanisms in the present case, though there are many examples where such data are ambiguous. Determination of the exchange rates at different pressures, however, produces a very convincing demonstration of the reality of this difference. Some spectra are shown in Fig. 6.15, together with calculated envelopes, from which the exchange rates were obtained. All the spectra are on the slow exchange side of coalescence, but it will be evident that the spectra of TMU are nearest coalescence at the lowest pressure, whereas the opposite is true for the DMSO. This difference

Table 6.1   Thermodynamic data for solvent S exchange on BeS_4^{2+}

Ligand	Rate at 298 K	ΔH^* (kJ mol^{-1})	ΔS^* (JK^{-1} mol^{-1})	ΔV^* (cm^3 mol^{-1})
DMSO	$213 \, m^{-1} s^{-1}$	35.0	-83.0	-2.5
TMU	$1.0 \, s^{-1}$	79.6	$+22.3$	$+10.5$

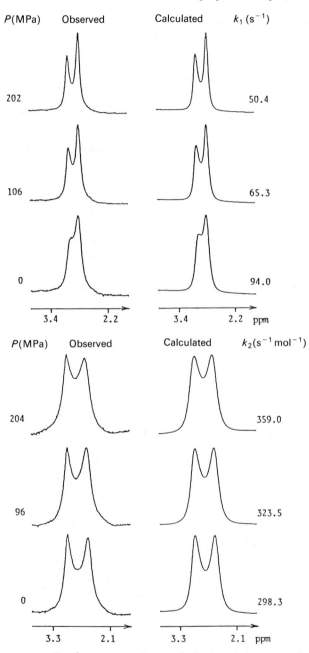

Fig. 6.15   The 200 MHz ^{1}H spectra of acetonitrile-d_6 solutions containing Be(II) and TMU (upper) or DMSO (lower), obtained as a function of pressure. The actual spectra are on the left and the calculated envelopes are on the right. (From Merbach (1987) *Pure Appl. Chem.*, **59**, 161, with permission.)

occurs because a dissociation produces a temporary increase in volume, which is discouraged by an increase in pressure, whereas association produces a temporary decrease in total system volume. The variation of the rate constant with pressure allows a volume change of activation to be obtained from

$$\Delta V^* = -RT(\delta \ln k/\delta P)_T$$

These values are also given in Table 6.1. ΔV^* is large and positive for TMU, and this is in accord with a fully dissociative mechanism for this ligand exchange. ΔV^* is negative for DMSO but its value is appreciably smaller, and this allows us to conclude that, while the mechanism is associative, there must be some tendency for the bound ligand to move away from the cation and that the mechanism is, more precisely, association and interchange of ligands near the cation, which limits the extent of the possible volume contraction were full association to occur. Note also the difference between the meaning of volume changes in this and the previous example. Here, the total volume before and after reaction is unchanged, and we observe only the adjustment necessary to allow reaction to proceed. There is thus no change in the position of equilibrium with pressure, in contrast to what happens when reaction produces a permanent volume change.

6.2.1   A technological application: crystallization of amorphous polyethylene

The proton spin relaxation time in amorphous, solid polyethylene, while shorter than that in liquids, is nevertheless long enough to obtain spectra by normal methods. When the material forms extended-chain crystals, the relaxation time becomes much shorter because the chain motion becomes more restricted, and this difference can be used to separate the signals of the two forms of the polymer. The technique used to do this was to apply the pulse sequence used to determine T_2, i.e. a $90°-\tau-180°-\tau-$echo sequence in which τ is made equal to $300\,\mu s$, too long to observe a response from the crystalline material but short enough to see an echo from the amorphous component. The echo intensity decreases with the time of crystallization as the amount of amorphous materials decreases. There is an induction period when no change in echo intensity is detected, and then the intensity falls regularly as crystallization proceeds. The rate increases as the temperature is decreased, and increased pressure allows crystallization to occur both at higher temperatures and at higher rates.

6.3   Reaction monitoring of systems not at equilibrium

The previous example was just such an experiment, in which we were monitoring a physical change in a system, albeit under somewhat extreme conditions. In fact, NMR is used extensively to monitor what is happening during chemical preparations and is particularly useful in that, because starting materials, intermediate, side products and main products may have resolved signals, then the progress of a reaction may well be fruitfully examined without any previous purification.

The slow equilibration of mixtures is easily followed by NMR. For instance, if the aluminate salts $NaAlMe_4$ and $NaAlEt_4$ are dissolved together in solution, it is found that the ligands scramble to form the mixed salts $NaAlMe_nEt_{4-n}$. If a coordinating solvent is used, the solutes ionize to a greater or lesser extent to give Na^+ and the regularly tetrahedral AlR_4^-, in which the ^{27}Al relaxation time can be sufficiently long to permit spin–spin coupling to the protons of the alkyl groups. In order to measure the rate of alkyl interchange, the two salts were dissolved in the non-coordinating solvent benzene and the coordinating solvent hexamethylphosphorotriamide (HMPTA), $(Me_2N)_3PO$, added in varying amounts. Immediately following dissolution, a sample of the solution was placed in an NMR spectrometer and the proton spectrum of the Al–Me groups measured as a function of time (Fig. 6.16). Any effect of the ^{27}Al was removed by double irradiation. The Al–Me signal then appears as a sharp singlet initially, but with time new signals are seen to emerge to high field corresponding to anions with one, two or three ethyl groups replacing the

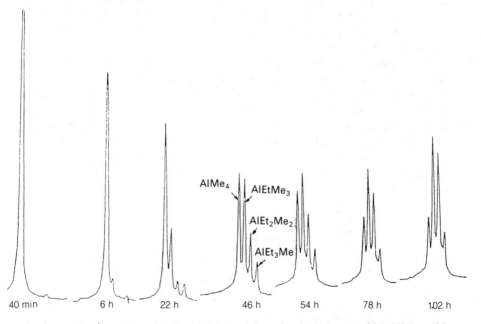

Fig. 6.16   The ^1H spectra of the Al–Me protons of a 1:1 mixture of $NaAlMe_4$ with $NaAlEt_4$ in benzene solvent with added HMPTA, as a function of time. Any effect of coupling to the ^{27}Al nuclei is removed by double resonance. (From Ahmad *et al.* (1984) *Organometallics*, **3**, 389; copyright (1984) American Chemical Society, reprinted with permission.)

methyl substituents. By measuring the intensities of the resonances as a function of time, the rate of ligand scrambling is obtained. Interestingly, this is found to be a function of the concentration ratio of HMPTA/Na$^+$, and it is believed that this indicates that the scrambling takes place in aggregates containing two anions and sodium cations solvated by the HMPTA.

Reactions taking as little as 60 s can be studied in this way, and even faster processes can be followed using stopped-flow NMR. Two solutions containing the species it is proposed to react are held in reservoirs in the magnetic field so that the magnetization of the nuclei to be studied can reach its equilibrium value. Two power-driven syringes are then used to force measured amounts of the solutions into the NMR sample tube within a time of a few milliseconds. The collection of FID signals is initiated at the same time and each response stored separately in memory for later transformation into spectra. Reactions taking as little as 5 s can be studied in this way, one example having already been given for the exchange of water on Al^{3+}.

Our next examples illustrate two reaction processes and how NMR can provide details about reactions not easily obtainable in other ways. The first example comes from the realm of organometallic chemistry. An oxidative addition was carried out on the platinum complex PtMe$_2$[Me$_2$PC$_6$H$_3$(OMe)$_2$]$_2$ using (bromomethyl) benzene (benzyl bromide), with the objective of forming the octahedral complex PtMe$_2$[Me$_2$PC$_6$H$_3$(OMe)$_2$]$_2$(PhCH$_2$)Br, where Ph is a phenyl group. The proton-decoupled ^{31}P spectrum of the starting material is shown in Fig. 6.17 and is the typical 1:4:1 triplet. The initial product has a similar spectrum, though with different shift and coupling constant but, even after 1.5 h reaction, there is already a significant amount of another product present. After three days this becomes a predominant spectral feature and is seen to consist of a 1:4:1 set of AB sub-spectra. The initial product has isomerized to give non-equivalent phosphines, which have a small *cis* P–P coupling. The full structure can then be deduced from the methyl resonance pattern in the ^{1}H spectrum, where one large and one small coupling to ^{31}P are observed, so indicating the presence of a *trans* and a *cis* methyl group. The reaction that takes place is:

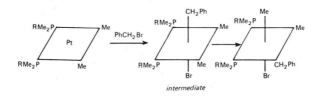

intermediate

where R = C$_6$H$_3$(OMe$_2$)$_2$ and Ph = phenyl.

The second example examines the nature of the species contained in highly hydrolysed aluminium salt solutions. Aluminium salt solutions are highly acidic and will, for instance, dissolve many metals with the evolution of hydrogen. If we add sodium carbonate solution slowly so as to ensure that there is no

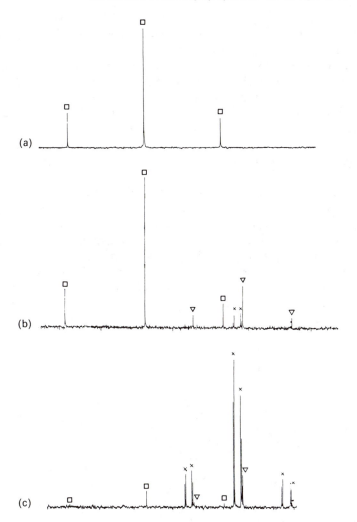

Fig. 6.17   Three 40.5 MHz ^{31}P spectra obtained with double irradiation of the protons so as to simplify the spectra. The upper spectrum (a) is that of the starting material and is distinguished by squares in all three spectra. The central trace (b) was obtained 1.5 h later and shows the presence of another substance in which the two phosphorus nuclei are still equivalent and which is distinguished by triangles. After three days reaction (c) both these spectral patterns are much diminished and a third, more complex, pattern (×) has taken their place. There is now a triplet of AB patterns, which shows that the two phosphorus nuclei in the complex are no longer equivalent. (Example supplied by Professor B. L. Shaw.)

precipitate formed, we find that we can add up to 1.25 moles per mole of Al^{3+} and form a solution whose stoichiometric composition is $Al(OH)_{2.5}X_{0.5}$, where X is the anion present. The rest of X forms the sodium salt and CO_2 is evolved. The problem has been to discover what exactly is present in such solutions, and ^{27}Al NMR has been able to provide useful new information about the nature of the molecular species formed. It was known that a cation AlO_4Al_{12} $(OH)_{24}(H_2O)_{12}^{7+}$ could be crystallized as the sulphate salt from the solutions to which the maximum amount of carbonate had been added. The structure of this cation has been obtained. It has a central aluminium atom tetrahedrally coordinated by four oxygen atoms and surrounded by 12 aluminium atoms, which are octahedrally coordinated and are linked by OH bridges and the oxygen atoms coordinated to the central Al. They also carry a terminal water ligand (Fig. 6.18). The four-coordinate aluminium atom is in a regular environment and so gives a narrow ^{27}Al resonance, which is diagnostic of the presence of this molecular species. The octahedral aluminium atoms are in a highly distorted environment, and so have a short relaxation time and a very broad resonance some 8000 Hz wide, which can only be observed under rather specialized conditions and so is normally not detectable. The system has been much investigated by pH titration methods but does not seem to be very amenable to these methods, although different workers had proposed the

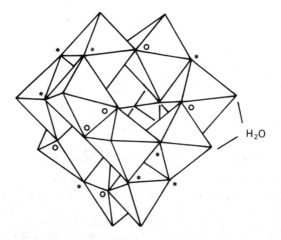

Fig. 6.18 Structure of the tridecameric hydrolysed aluminium cation, $AlO_4Al_{12}(OH)_{24}$-$(H_2O)_{12}^{7+}$. This consists of four groups of three AlO_6 octahedra, which share an apical oxygen atom and three edges to form a triangular structure. The apical oxygen is also co-ordinated to the central Al atom, which is thus in a tetrahedral environment since there are four Al_3 groups. Each Al_3 group is attached to the three others via double OH bridges marked (O) on the figure. There are also three OH bridges in each triangular cluster (∗). The remaining oxygen coordinated to the Al is fully protonated and is a water ligand. The central Al is not shown. (From Akitt and Elders (1988) *J. Chem. Soc. Dalton Trans.*, 1347, with permission.)

presence of some 24 total species in order to explain their results. Later workers, however, seemed to be agreed that only the tridecameric cation was present. Unfortunately, in partly neutralized solutions, the ^{27}Al NMR gives a spectrum containing three resonances, none very broad. These can be seen in the traces of Fig. 6.19, where the lowest-field line is assigned to the tridecamer, the highest-field line to unconverted Al^{3+} or $Al(H_2O)_6^{3+}$ and a rather broader line just to low field of the latter is labelled as an oligomer. There is no doubt that this species exists but it is not detected in the pH titrations. A major difference between the two techniques is that pH titrations are carried out in rather dilute solutions whereas, because NMR is a rather insensitive technique, it requires much higher concentrations. (This difference is currently less important because of the existence of the very high-field spectrometers coupled with Fourier transform techniques.) The obvious question was asked: Was this difference in interpretation a concentration effect? It was found that, if a solution that had

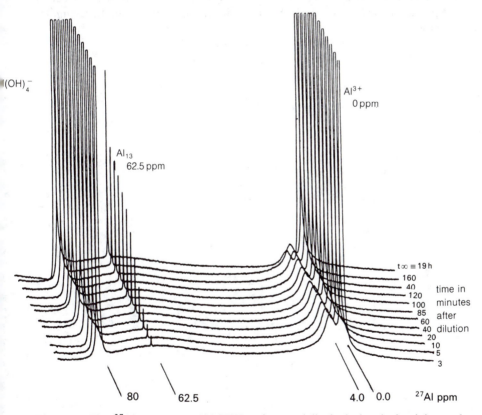

Fig. 6.19   The ^{27}Al spectra at 104.2 HHz of a partially hydrolysed aluminium salt solution as a function of time immediately following dilution with water. The aluminium concentration before dilution was 0.5 M and this was reduced to 0.05 M by dilution. (From Akitt and Elders (1988) *J. Chem. Soc., Dalton Trans.*, 1347, with permission.)

been hydrolysed so as to contain very little tridecamer and much oligomer was diluted, then the relative concentration of the tridecamer increased and that of the oligomer decreased, and that, with sufficient dilution, only the tridecamer could be detected. It was also possible to follow how the change occurred by simply diluting a solution with water and observing the ^{27}Al resonance as a function of time. A set of spectra are shown in Fig. 6.19. The oligomer resonance is reduced in intensity immediately upon dilution – a change that cannot be followed by this technique and which requires comparison of the spectra before and immediately following dilution – and the tridecamer concentration then increases regularly over about 19 h. Simple dilution of these solutions, then, causes profound changes in composition and this explains some of the problems of the earlier workers. It is also remarkable that such a weak driving force as a quite moderate change in concentration should give rise to such a complex molecule as the tridecamer. It is suggested that the oligomer is, in fact, a mixture of species that have structures related to those of fragments of the tridecamer, one possibility being a fused $Al_3O_{13}H_{18}^+$ unit such as forms one of the triangular, flat faces of the tridecamer.

7 Multiple resonance and the nuclear Overhauser effect

We have already seen in the last chapter that it is possible to subject the nuclei in a sample to several radiofrequency fields. In general, the signal obtained from the nuclei that are being observed will be influenced in some way if other nuclei in the same molecule are irradiated at the same time, and in certain cases intermolecular effects can be observed. This is called the double-resonance experiment, and it is used either to simplify the spectrum of a particular nucleus or to probe correlations between different nuclei such as spin–spin coupling or through-space cross-relaxation occurring via the dipole–dipole mechanism. It is possible to use more than one perturbing frequency and we can have triple, quadruple, ... resonance experiments, though the experimental difficulties associated with such experiments are not trivial and the more complex situations seem to be better tackled by two-dimensional techniques, which we will discuss in Chapter 8. It is normal to distinguish two different types of experiment: homonuclear double resonance, where the nuclei irradiated are the same isotope as those observed – observing and irradiating protons, for example, for which the short-hand notation is $^1H\{^1H\}$; and heteronuclear double resonance, where the nuclei irradiated are different from those observed – observing ^{13}C while irradiating ^1H, for example, for which the short-hand notation is $^{13}C\{^1H\}$. In other words, the second irradiation is applied to the nucleus in the braces. This irradiation is called B_2 and can be applied in many different ways depending upon what effects it is wished to study. Double irradiation was first applied to the study of continuous-wave (CW) spectra and to probing the coupling paths between the multiplets observed, which is a very useful way of assigning the resonances of complex spectra.

7.1 Complete decoupling

If we consider a nuclear system A_nX_m in which all nuclei A are equivalent as are all nuclei X, and the two types of nuclei are spin-coupled, then the resonances of X and A will both be split into multiplets whose multiplicity depends upon the values of n and m and the spin quantum numbers of A and X. A and X may be the same isotope and so chemically shifted, preferably by a substantial amount, or they may be different isotopes and so with markedly different NMR

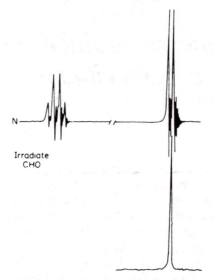

Fig. 7.1   An example of high-power homonuclear double irradiation. N shows the normal spectrum of acetaldehyde, CH_3CHO, and below it the spectrum of the methyl group run while the formyl proton was double-irradiated. (From Bovey (1965) *Chem. Eng. News*, 30 Aug., p. 118; copyright (1965) American Chemical Society, reprinted with permission.)

frequencies. We will observe A while applying a strong radiofrequency field B_2 to X, remembering that in the homonuclear case this may require some technical modifications of the experiment. B_2 has to have the same frequency as the centre of the X multiplet and so is stationary, or nearly so, in the rotating frame relative to these nuclei. The X nuclei then precess around B_2 at a frequency that depends upon the magnitude of B_2, and, if this precession is fast enough, their z magnetization effectively disappears at the coupled A nucleus whose multiplicity disappears and so becomes a singlet. A is said to be decoupled from X. For decoupling to be complete, it is necessary that

$$\gamma_X B_2 / 2\pi \gg J(AX)$$

If B_2 is small, then the A spectra can become more complex and exhibit extra splittings. Two examples of complete decoupling are shown in Figs 7.1 and 7.2, both being run in the CW mode. The chemical shift between the two proton resonances of acetaldehyde is almost 10 ppm and the coupling constant is small, so that this homonuclear decoupling is relatively facile. Note that the formyl proton quartet is not shown in the decoupled trace as B_2 interferes strongly with B_1. Decoupling is established in such an experiment very quickly since precession around B_2 starts as soon as B_2 is a applied.

At the same time as the precession of individual nuclei takes place under the influence of B_2, there is saturation of the X spin system. The word 'saturation' derives from usage in CW spectroscopy, where it is found that, if the B_1

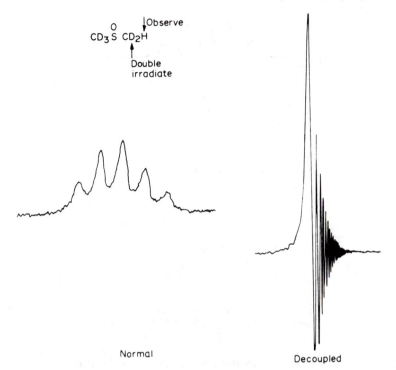

Fig. 7.2   The ^{1}H spectrum of pentadeuteriodimethylsulphoxide. The proton resonance of the CD$_2$H group is split into a quintet by the two deuterons. The splitting is removed when the deuterium resonance is strongly irradiated. (Spectrum supplied by Bruker Ltd.)

power is too large, the intensity of the signal decreases. The spin system absorbs energy from the B_1 field so that the low-energy excess of spins becomes depleted, a process that is in competition with the T_1 relaxation processes. This also happens for B_2, and if this is large enough then the spin populations become equalized and the total magnetization M_z become zero. This is a non-equilibrium state and all the relaxation mechanisms present will work towards the re-establishment of equilibrium. If the predominant mechanism of relaxation is dipole–dipole, then the energy exchange required by this relaxation causes mutual spin flips of A and X, and the spin population of the A nuclei is disturbed, so that decoupling between spin-1/2 nuclei is usually accompanied by intensity modifications of the signal of the decoupled nucleus. This is the nuclear Overhauser effect (NOE). Because relaxation processes are involved in determining the NOE, this is established much more slowly than the decoupling.

   In more complex molecules, there may be two or more types of X nuclei coupled to the A nuclei, and, provided the chemical shift between the types of X nuclei is larger than any coupling between them, then it is possible to carry out selective decoupling experiments in which the effects of each group of X

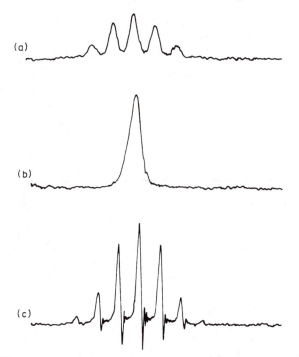

Fig. 7.3   The ^{31}P spectrum of triethyl phosphite, $(CH_3CH_2O)_3P$: (a) obtained without double irradiation, showing coupling to the methylene protons broadened by the coupling to the methyl protons; (b) the methylene protons irradiated, which selectively removes their influence from the spectrum; (c) the methyl protons irradiated, which gives a well resolved septet due to coupling to the methylene protons. (Spectrum supplied by Bruker Ltd.)

nuclei can be examined in turn. A heteronuclear example is that of the ^{31}P spectrum of triethyl phosphite, $(CH_3CH_2O)_3P$, whose spectra are shown in Fig. 7.3. The phosphorus nucleus is coupled to both types of proton, that to the methylene protons giving a septet and that to the methyl protons giving a decet. The coupling to the methyl protons is, however, quite small, and the normal spectrum is a septet of broad lines that contain the decet structure due to the coupling to the methyl group. This broadening reduces the intensity of the resonances, to the extent that it is difficult to observe the two weak outer lines of the septet (Fig. 7.3(a)). Because the proton resonances of the ethyl and ethylene groups are separated by much more than the interproton coupling constant, it is possible to irradiate one with B_2 sufficient to eliminate the coupling to the ^{31}P without affecting the coupling of the other type of protons. One effectively eliminates one of the coupling interactions, observes what the other produces and then irradiates the other group to observe the effect of the first alone. The $P(CCH_3)_3$ fragment spectrum is in Fig. 7.3(b) and the well resolved methylene coupling in Fig. 7.3(c). It is, of course, possible in principle to apply sufficient B_2 power that both types of proton are decoupled from the phosphorus, whose resonance then becomes a singlet. In practice, it proves technically difficult to provide sufficient power to do this, and means have to be found to overcome this difficulty. The use of two separate B_2 frequencies at the same time is possible, but it has been found that the best method is to ensure a broad band of frequencies throughout, in this case, the proton chemical shift range. This is achieved by frequency modulation of the B_2 frequency with a randomly varying waveform or by using sequences of short pulses of energy to swing the spins around. With such a spread of frequencies, the power requirements are much less stringent. The technique is called broad-band decoupling and is used extensively, indeed almost invariably, when observing the nucleus ^{13}C in hydrogen-containing compounds. The hydrogen is broad-band irradiated and this removes all spin coupling to the carbon atoms, so simplifying their resonances to narrow singlets, increasing the intensity of the resonance of a given type of carbon atom and so reducing markedly the time necessary to obtain a given signal-to-noise ratio with this insensitive nucleus. In addition, there is also a valuable intensity enhancement due to the Overhauser effect, which combined with the simplification of the spectra gives a time reduction factor of the order of 250 times compared with a spectrum obtained without double irradiation.

7.2   The nuclear Overhauser effect

In Fig. 7.4 we show an energy level diagram for a spin-1/2 nucleus I. The number of nuclei in the upper energy state, N_u^I, is less than those in the lower energy state, N_l^I, and this state is maintained despite the existence of transitions between the two spin states, since the probabilities of upwards and downwards transitions are different. We designate these W_+ and W_- respectively. We have from section 1.3 that the ratio between the numbers in the upper and lower energy states is

$$N_u^I/N_l^I = W_+^I/W_-^I = \exp(-\gamma_I \hbar B_0/kT) \tag{7.1}$$

We will suppose that these spin transitions, and so relaxation, are caused entirely by the spins S of a nucleus that is either a different species or the same but chemically shifted significantly from the spin I. The influence of S occurs directly through space; spin–spin coupling may or may not be present between I and S, and is in any case a much weaker interaction. The through-space interaction is, of course, modulated by the motion of the molecule at correlation time τ_c, which continuously changes the direction of the I–S vector relative to B_0. At equilibrium, this process maintains the relative spin I populations. From equation (7.1), remembering that $e^x = 1 + x$ is a good approximation, we can obtain the fractional difference in populations between the two energy states:

$$(N_l^I - N_u^I)/N_l^I = \gamma_I \hbar B_0/kT \tag{7.2}$$

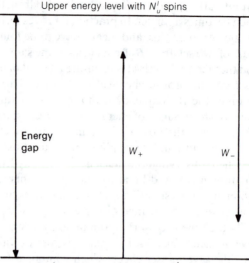

Upper energy level with N_u^I spins

Energy gap W_+ W_-

Lower energy level with N_l^I spins

Fig. 7.4   Energy level diagram for a spin-1/2 nucleus in a magnetic field $\boldsymbol{B}_0$. W_+ and W_- are the transition probabilities for upward and downward transitions, respectively. They are not equal, since the population of the lower energy state is greater than that of the upper.

and a similar expression can be obtained for the difference in the two S spin populations

$$(N_l^S - N_u^S)/N_l^S = \gamma_S \hbar B_0/kT \tag{7.3}$$

The signal intensity obtained from the I nuclei is proportional to the fractional population difference described by equation (7.2). Thus the Boltzmann equilibrium of spins I is maintained by the Boltzmann equilibrium of spins S. If S is strongly irradiated, then it is saturated, and its spin populations are perturbed and equalized. S is no longer at its Boltzmann equilibrium, and so it cannot maintain the Boltzmann equilibrium of spins I, and the intensity of the I signal is changed by the irradiation of spins S. To equalize the S populations, it follows from equation (7.3) that there is a population change of $\pm \frac{1}{2}\gamma_S \hbar B_0/kT$ for the two energy levels of S. This will produce a proportional change in the I populations, which, if the constant of proportionality is ϕ, becomes $\gamma_I \hbar B_0/kT + \phi \gamma_S \hbar B_0/kt$. Thus we have the intensity enhancement

$$\frac{\text{intensity } I \text{ with } S \text{ irradiated}}{\text{normal intensity } I} = \frac{\gamma_I \hbar B_0/kT + \phi \gamma_S \hbar B_0/kT}{\gamma_I \hbar B_0/kT}$$

$$= 1 + \phi \gamma_S/\gamma_I$$

$$= 1 + \eta_{IS} \tag{7.4}$$

where η_{IS} is the nuclear Overhauser enhancement factor, usually expressed simply as η. Two factors contribute to η in equation (7.4): the ratio of the magnetogyric ratios of the two different spins and the factor ϕ. The latter, since it is related to relaxation processes, is a function of $\omega\tau_c$, and the behaviour is different inside and outside the extreme narrowing region. It is also a function of the exact relaxation mechanism. Provided this is purely dipole–dipole and we are working in the extreme narrowing region, then ϕ has the value 1/2. If other relaxation mechanisms affect spin I, for example chemical shift anisotropy relaxation, then ϕ is reduced and can become zero. No Overhauser effect is possible for a quadrupolar nucleus except for the very rare cases where a combination of low quadrupole moment and high symmetry renders the quadrupolar mechanism less important than the dipolar one. It is, indeed, possible to use measurements of η to calculate the proportion of the dipolar mechanism operating for a particular nucleus. The value of η given by $\gamma_S/2\gamma_I$ is then to be regarded as a maximum value, η_{max}. If the NOE is measured to be smaller, η, then.

$$\eta/\eta_{max} = T_1/T_{1DD}$$

where T_1 is the measured relaxation time of I and T_{1DD} is the pure dipolar relaxation time. We note also that dipole–dipole relaxation caused by more distant, non-irradiated nuclei does not contribute to T_{1DD} as used here, which relates to the irradiated nucleus only. Fortunately, the r^{-6} dependence of relaxation means that such interactions can often be ignored. The value of ϕ remains 1/2 throughout the extreme narrowing region but falls off in the region of the T_1 minimum, defined in Fig. 4.4, and is zero where the correlation times are longer still. NOE effects may not be possible for large molecules at high magnetic fields.

The other factor that determines η is the ratio γ_S/γ_I. In the homonuclear case, these two are equal and η_{max} is 1/2. Such is the value when probing intra-proton interactions. The effect is then small and special techniques are needed to ensure that a true NOE is being observed. In other cases, it is obviously necessary to choose pairs of nuclei where $\gamma_S > \gamma_I$ so as to ensure that the NOE is significant. For the observation of ^{13}C, for instance, when protons in the molecule are double-irradiated, the ratio is 1.99 and $1 + \eta_{max}$ is effectively 3. This is a very useful gain in intensity for such low receptivity, and explains why ^{13}C spectra are routinely obtained with proton broad-band irradiation. Some values of η_{max} are shown in Table 7.1. We should note that some nuclei have magnetogyric ratios with negative sign and that η_{max} is then negative also. In such cases, provided η_{max} is more negative than -1 (^{15}N and ^{29}Si in the table), then the signal is inverted relative to the normal one and the total enhancement is less than η_{max}. If the relaxation mechanism contains other contributions than the direct dipolar one, then η_{max} will be reduced, and it is not uncommon to find that $1 + \eta_{max} = 0$ and all signal is lost. In such cases, it is necessary to suppress the NOE in some way, and this can be achieved either instrumentally

Table 7.1   *Maximum nuclear Overhauser effects for several pairs of nuclei*

Irradiate	^{1}H						^{19}F		
Observe	^{1}H	^{13}C	^{15}N	^{19}F	^{29}Si	^{31}P	^{1}H	^{13}C	^{19}F
η_{max}	0.5	1.99	−4.93	0.53	−2.52	1.24	0.47	1.87	0.5
$1 + \eta_{max}$	1.5	2.99	−3.93	1.53	−1.52	2.24	1.47	2.87	1.5

or by adding a paramagnetic salt (a relaxation agent), which completely dominates the relaxation processes.

7.3   Application to ^{13}C spectroscopy

Figure 7.5 illustrates the changes brought about in the ^{13}C spectrum of ethylbenzene by broad-band proton double irradiation. In the absence of irradiation, all the ^{13}C–^{1}H spin coupling interactions are observed. Thus the protons of the phenyl group, seen to low field, are split into doublets by the proton bonded directly to the carbon atom, and these lines are further split by longer-range couplings to the other ring protons. Only the quaternary carbon atom gives an apparent singlet, though there will be some long-range splitting. A triplet is seen in the centre of the trace due to the solvent CDCl$_3$ whose ^{13}C resonance is coupled to the ^{2}D nucleus. The CH$_2$ and CH$_3$ resonances are to high field and are triplet and quartet, respectively, due to the directly bonded hydrogen nuclei, and with longer-range two-bond coupling to the vicinal hydrogens, which produces fine structure in the resonances. Not all the resonances of these multiplets can be seen in the spectrum illustrated as the conditions were chosen to produce a noisy baseline. The difference in intensity between the phenyl quaternary carbon signal and the remaining phenyl carbon signals is due to the difference in relaxation time. That of the quaternary carbon atom is long and, at the pulse repetition rate used to obtain the spectrum, has insufficient time to recover its z magnetization after each pulse and so its signal is reduced in amplitude. The spectrum is much simplified by double irradiation, and each type of carbon nucleus appears as a singlet. The same conditions were used to obtain the two spectra illustrated and the enormous improvement in the signal-to-noise ratio is immediately evident. The line intensities, starting at the quaternary carbon resonance, should be in the ratio 1:2:2:1:1:1, but clearly differ quite seriously from these figures. This is as a result of both relaxation time differences and of different NOEs. The quaternary carbon resonance is even weaker relative to the other resonances than it was without double irradiation, and this is because, being further from the protons than the other carbon atoms, it has a smaller NOE. Similarly, the solvent resonance has no NOE and is not detected when the irradiation is applied.

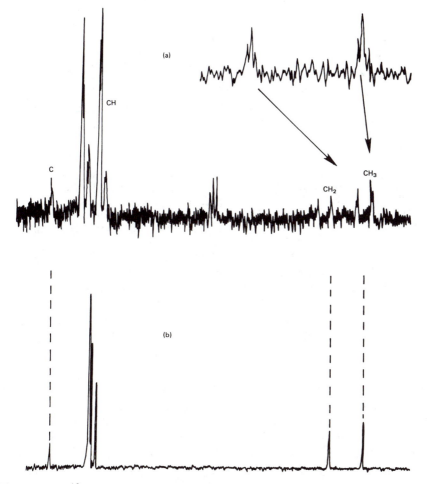

Fig. 7.5   The ^{13}C spectra at 22.6 MHz of ethylbenzene, $C_6H_5CH_2CH_3$, dissolved in deuteriochloroform: (a) obtained without proton double irradiation and (b) with irradiation at 90 MHz, otherwise the spectrometer conditions were identical. A large improvement in signal-to-noise ratio is obtained with double irradiation.

Double irradiation thus permits us to obtain ^{13}C spectra at natural abundance routinely with good signal-to-noise ratio, but such spectra contain reliably only the chemical shift information. Often, this is sufficient, but there are instances where it may be useful to observe the coupling patterns, and if this could be done while retaining the NOE then much accumulation time could be saved. Alternatively, it might be useful to know the correct intensities, and for this we need to decouple but suppress the NOE. Such techniques may be necessary if we are to observe quaternary carbon atoms with very long relaxation times. If we are happy to contaminate the sample, then the relaxation agents can be

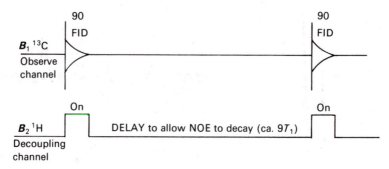

Fig. 7.6   Decoupler timing to allow a fully decoupled spectrum to be obtained without any distortion of intensity due to the NOE. There will be a small build-up of NOE during the B_2 pulse (typically 0.5 s) and this must be allowed to die away completely before the next pulse. The long delay time means also that the nuclear magnetization has decayed fully before the next 90° pulse and there are no intensity distortions due to relaxation effects.

very useful since these effectively suppress the NOE and also shorten the ^{13}C relaxation times, so permitting a shorter delay between pulses. If contamination cannot be tolerated, or if the natural parameters of the molecule are required, then special pulse sequences are used, which depend upon the different timescales needed to set up decoupling and NOE. Thus we can have the following:

1. *Full decoupling without NOE.* This is achieved by decoupling only for the short time needed to collect the FID and then waiting sufficient time to allow any small NOE population changes to return to normal. The short pulse of B_2 does not allow appreciable build-up of the NOE (Fig. 7.6).
2. *No decoupling but full NOE.* Here the decoupling power is left on for sufficient time for the NOE to build up to its full value and is then switched off while the FID is collected (Fig. 7.7). This technique allows normal spectra to be obtained more quickly. If bad overlap of the multiplets occurs, a different

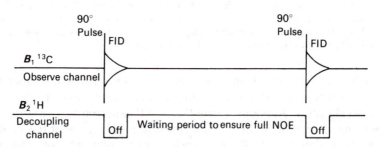

Fig. 7.7   Decoupler timing to allow a non-decoupled spectrum to be obtained but with the benefit of the full NOE increase in intensities. There will be very little fall-off in NOE during the short time needed to collect the FID data. This technique allows up to nine times reduction in time over the basic method used to obtain a non-decoupled spectrum.

approach may be needed, but normally it is possible to distinguish quartets, triplets, doublets and singlets by a comparison with the decoupled spectrum and so to assign resonances to CH_3, CH_2, CH and quaternary carbons. It is well worth taking the trouble to carry out this experiment because the NOE enhancement permits a time reduction in accumulation of as much as nine times.

A variety of other experiments is also possible. Off-resonance decoupling will give multiplet structure with apparently reduced coupling constants, which will reduce overlap in crowded spectra and permit assignment of the carbon spectrum and its correlation with the proton spectrum. Such spectra obtained at several spot frequencies will indicate at which ^1H frequency a given ^{13}C multiplet becomes a singlet, and so permit an even more precise correlation between the two. Such experiments are, however, time-consuming and are now better carried out by two-dimensional techniques.

7.4 Detailed relaxation mechanisms by ^{13}C NOE and T_1 measurements

The relaxation time of ^{13}C nuclei can be obtained by the inversion-recovery method outlined in section 5.13, though this is not the only method available. The NOE factors η are obtained by comparing integrals with and without double irradiation, and if the relaxation mechanism is dipole–dipole then η should have a value near to 2. If η is less than this, then other mechanisms are present. An alternative method of measurement of T_1 and NOE in the same experiment is also possible and is done by a method that is a combination of the two previous ones. The NOE is allowed to build up for a time t prior to producing the FID and this is collected while irradiation continues. B_2 is then switched off, the system allowed to equilibrate and the process repeated. A series of experiments are run with different values of t (Fig. 7.8). Such experiments are known as dynamic NOE measurements and give both the Overhauser enhancement for each nucleus from the intensity of the signals when $t = 0$ and $t = \infty$, and the value of T_1 from the plot of the change in signal intensity as a function of t. The spectra obtained for such an experiment with biphenyl are shown in Fig. 7.9, and the results for biphenyl and some for toluene are illustrated in Table 7.2. The CH ring carbon atoms of biphenyl have an NOE only slightly less than 3 and so must be almost totally relaxed by the directly bonded hydrogen atoms. The carbon nuclei 2 and 3 have similar relaxation times but the T_1 of carbon 4 is appreciably shorter, and this is because the CH bond is not reoriented relative to B_0 by rotations around the axis of the molecule, whereas this motion reduces the correlation time of the carbon atoms 2 and 3, if only slightly. Much more marked is the long T_1 of the quaternary carbon nucleus, which relies on the long-distance effect of the protons. Its NOE, however, is reduced and indicates that only half the rate of relaxation is due to this dipole–dipole

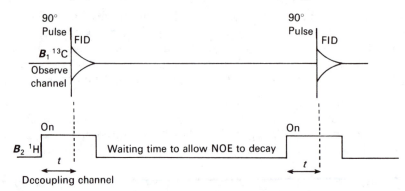

Fig. 7.8   The dynamic NOE experiment. This is essentially a combination of the two experiments described by Figs 7.6 and 7.7. The decoupler is first gated ON for time t to allow the NOE to build up. The amount of NOE increases if t is increased, reaching a maximum when $t > 9T_1$. At the end of time t the FID is produced by the 90° pulse and collected with continuing irradiation. This is removed when the data collection is finished and the NOE allowed to decay to zero. Sufficient FIDs are collected to give the required signal-to-noise ratio, and the experiment is repeated with different values of t, but always the same number of FIDs are collected, so that spectral intensities can be compared directly.

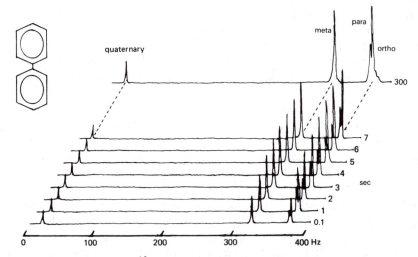

Fig. 7.9   Fourier transform ^{13}C spectra of biphenyl in $CDCl_3$ solution excited by 90° read pulses separated by intervals of 300 s in order to ensure the re-attainment of equilibrium after each pulse. The spectra are stacked as a function of the time t of proton irradiation prior to the application of the pulse. (From Freeman *et al.* (1972) *J. Magn. Reson.*, **7**, 327, with permission.)

Table 7.2   Experimental spin–lattice relaxation times and nuclear Overhauser enhancements for ^{13}C nuclei in biphenyl and toluene

(a) Biphenyl

Carbon atom	T_1 (s)	$I_\infty/I_0 = 1 + \eta$	η
C4	3.4 ± 0.6	2.72 ± 0.16	1.72
C3	5.4 ± 0.7	2.80 ± 0.21	1.80
C2	5.2 ± 0.8	2.76 ± 0.19	1.76
C1	54 ± 0.4	2.00 ± 0.16	1.00

(b) Toluene

Carbon atom	T_1 (s)	$I_\infty/I_0 = 1 + \eta$	η
CH_3	$16\ (T_1 N_H = 48)$	1.61	0.61
C1	89		
C2	24		
C3	24		
C4	17		

interaction and that another influence is operating, which we can suggest is that of chemical shift anisotropy. The T_1 values for the ring carbon atoms of toluene show similar behaviour, though the actual relaxation times are longer, as would be expected for a smaller, more rapidly rotating molecule. The relaxation time of the methyl carbon nucleus has, of course, to be multiplied by 3, the number of directly bonded protons, N_H, if it is to be meaningfully compared with the T_1 values of the CH carbon nuclei, and this gives a value of 48 s. This is really very long when compared with the ring CH carbons, and it is not surprising to observe that the NOE is also small, with only 30% of the relaxation rate arising from the dipolar mechanism. In this case, the dipolar effect must be reduced because of a very short correlation time, much shorter than for the rest of the molecule. This will be the free, almost unhindered, rotation of the methyl group, which in addition causes spin rotation relaxation and so accounts for the small NOE. One can calculate that the part of the relaxation which is purely dipolar, $3T_{1DD}$, has the value 160 s, which implies a reduction of τ_c for the methyl group of around eight times compared with the rest of the molecule.

A more complex and more informative example concerns the relaxation behaviour of the carbon nuclei in the pyridinepentacarbonylchromium complex, $C_5H_5N–Cr(CO)_5$. This is a σ-bonded complex in which one CO ligand in $Cr(CO)_6$ has been replaced by an N-bonded pyridine molecule. The chemical interest in such molecules arises because it is uncertain what sort of energy barriers exist to the rotation of the $Cr(CO)_5$ moiety relative to the aromatic ligand. One way of probing the intramolecular motion is to measure the relaxation rates of the ^{13}C nuclei. Provided the mechanism of relaxation is known, then correlation times of individual atoms can be calculated. Two types of mechanism are expected to coexist in such molecules: dipole–dipole for the protonated carbon atoms and chemical shift anisotropy relaxation for the CO carbon atoms. The first mechanism can be confirmed if the NOE is high, and low NOE establishes its absence for the CO atoms. Another experiment is, however, needed if we are unequivocally to establish the presence of CSA relaxation, and this is achieved by measuring T_1 and η at different magnetic fields, though the rates of motion must be such that the extreme narrowing condition is met at the higher field and spectrometer frequency.

The T_1 values were in this case obtained by a saturation-recovery technique in which the ^{13}C spin populations are equalized by the application of a series of closely spaced 90° pulses and the recovery of the signal intensity monitored as a function of time. The NOE was determined by comparing signal intensities obtained with continuous decoupling with those using gated decoupling of the type depicted in Fig. 7.6. $CDCl_3$ was used as solvent and solutions were degassed to ensure that there was no contribution to relaxation from solvent nuclei or dissolved paramagnetic oxygen. The two spectrometer frequencies used were 125.7 MHz (11.7 T, ^1H resonates at 500 MHz) and 50.29 MHz (4.7 T, ^1H at 200 MHz). Certain assumptions had to be made in order to calculate the chemical shift anisotropy of the carbonyl carbon atoms. Since the *trans* CO is on the molecular axis, its relaxation is not affected by rotation around this axis since the CSA will be axially symmetric. The same comment applies to the relaxation of the γ carbon atom of the pyridine since axial rotation does not reorient the C–H bond. One can therefore assume that these two carbon atoms have the same correlation time of motion, that of the end-over-end tumbling of the complex. The correlation time for the γ carbon is then calculated from its T_{1DD} using equation (4.3) and this value is used to calculate $\sigma_\parallel - \sigma_\perp$ for CO using equation (4.6). The full results are shown in Table 7.3. First we see from the chemical shifts of the carbonyl carbons that the *cis* and *trans* carbonyl groups are very similar, so that we can assume similar chemical shift anisotropy in calculating the correlation times of the *cis* carbonyl carbons. The values of $T_1, R_1 (= 1/T_1)$ and η are given for each carbon at both magnetic fields. For the nuclei of the ring carbon atoms, the η values are quite substantial, and there is only a relatively small decrease in relaxation time with increase in magnetic field. The mechanism is predominantly dipolar, and we can extract R_{1DD} with reasonable precision, and so the correlation times. In the case of the carbonyl

Table 7.3   Relaxation and NOE data for the ^{13}C nuclei in two chromium carbonyl complexes

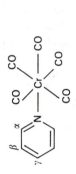

Pyridinepentacarbonylchromium η^6-Benzenetricarbonylchromium

(a) Pyridine complex

Atom	δ_C (ppm)	50.29 MHz			125.7 MHz					
		T_1 (s)	R_1 (s^{-1})	η	T_1 (s)	R_1 (s^{-1})	η	R_{1DD}	R_{1CSA}	τ_c (ps)
α	155.3	7.52	0.133	1.7	5.86	0.171	1.4	0.116	0.017[a]	5.5 ± 0.9
β	124.8	7.01	0.143	1.6	4.47	0.224	1.2	0.123	0.014[a]	6.0 ± 1.0
γ	137.1	2.44	0.410	1.6	2.16	0.463	1.3	0.319	0.102[a]	16.3 ± 2.8
trans-CO	220.7	10.3	0.097	0.0	2.63	0.380	0.0	0.000	0.338	assumed same as above
cis-CO	214.3	25.6	0.039	0.1	3.96	0.253	0.0	0.000	0.143	12.8 ± 2.8

(b) Benzene complex

Atom	δ_C (ppm)	50.29 MHz			125.7 MHz					
		T_1 (s)	R_1 (s^{-1})	η	T_1 (s)	R_1 (s^{-1})	η	R_{1DD}	R_{1CSA}	τ_c (ps)
CH	92.7	11.7	0.085	1.8	10.8	0.093	1.4	0.070	0.021[a]	3.6 ± 0.6
CO	232.8	31.3	0.032	0.1	6.61	0.151	0.1	0.005	0.142	6.3 ± 1.4

[a]These values probably embrace contributions from several mechanisms. R_{1DD} is calculated from the results at both frequencies and the average taken. The calculated $R_{1DD} + R_{1CSA}$ should equal R_1, found by experiment at the higher frequency, but the averaging process destroys this equality. The values calculated for R_{1CSA} apply to the higher frequency only. The accuracy of the experimental T_1 values is approximately ±7% and the error on η is ±0.2.

Source: Data adapted from Gryff-Keller *et al.* (1990) *Magn. Reson. Chem.*, **28**, 25; copyright (1990) John Wiley and Sons Ltd, reprinted with permission.

carbons, η is essentially zero, and there is little long-range dipolar interaction. T_1 is very field-dependent and T_{1CSA} can be calculated from the values obtained at each field and, thus, the correlation time of the *cis* carbonyls. It is evident that the rate of movement of the *cis* carbonyl groups is about half that of the reorientation of the plane of the pyridine ring, so that the two halves of the complex are rotating independently around the N–Cr bond, with the rates of motion determined primarily by interactions of the groups with the solvent. The data for the (η^6-arene) complex, η^6-$C_6H_6Cr(CO)_3$, are also given in Table 7.3. They resemble closely the previous set of data, though the correlation times calculated are much shorter and the speed of rotation of the benzene is high, indicating its very small interaction with the solvent.

7.5   Low-power double irradiation

If the magnitude of B_2 is small, then the frequency v_2 given by

$$v_2 = \gamma B_2/2\pi$$

is small, and the irradiation can perturb only a small part of a spectrum. This technique is used to investigate connections between different parts of a spectrum, or spectra in the heteronuclear case, and, while largely superseded by the 2D experiments for complex molecules, is still used for the simpler cases.

We will first consider how weak irradiation can affect an AX spectrum. The energy level diagram for this is shown in Fig. 7.10, where there are four energy levels depending upon the orientations of the two spins and increasing in energy from 1 to 4. The population differences depend upon both A and X population

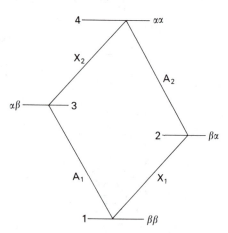

Fig. 7.10   An energy level transition diagram for a system of two coupled spin-1/2 nuclei, A and X.

differences, and, if these are N_a for levels 1, 3 and 2, 4 and N_x for 1, 2 and 3, 4, then the population differences at equilibrium can be written as supplements to a basic population of each state, N. Thus

$$\text{level 1 population} = N + \tfrac{1}{2}N_a + \tfrac{1}{2}N_x$$
$$\text{level 2 population} = N + \tfrac{1}{2}N_a - \tfrac{1}{2}N_x$$
$$\text{level 3 population} = N - \tfrac{1}{2}N_a + \tfrac{1}{2}N_x$$
$$\text{level 4 population} = N - \tfrac{1}{2}N_a - \tfrac{1}{2}N_x$$

If now we irradiate one of the transitions, say X_1, we will equalize the populations without modifying any of the others if B_2 is weak enough. N_x becomes zero for levels 1 and 2, and the population of level 1, which is the most highly populated, is reduced, so that the intensity of transition A_1 is reduced, by half in the homonuclear case. On the other hand, the population of level 2 is increased, and the intensity of transition A_2 is increased, the change again amounting to one-half. We note that X_2 is not affected by these changes and that it is only those transitions which have energy levels in common that show these changes. The experiment is known as an internuclear double resonance (INDOR) experiment and is used to probe the connections between lines in multiplets in spin-coupled spectra.

The method was developed for CW spectrometers but has been adapted to the FT instruments. First, a normal spectrum is obtained and stored in memory but with irradiation B_2 operating at a remote, non-interacting frequency. B_2 is then set on the resonance it is wished to irradiate and a spectrum run, accumulating an equal number of transients. The two spectra are then subtracted, when only the perturbed lines will show up, since these are the only ones whose intensity has been changed. An INDOR spectrum is shown in Fig. 7.11. B_2 is run continuously throughout because the spectrometer parameters may alter when it is switched on, and this would spoil the difference spectrum by introducing intensity from not quite cancelled strong resonances. This technique of difference spectroscopy is much encountered in FT NMR. In addition, since an FT spectrometer is a wide-band device, the B_2 frequency will be picked up by the receiver and will interfere with the response following the B_1 pulse. This can be avoided by using time-shared decoupling in which B_2 is only switched on while the receiver is off. This mode of operation is made possible because the receiver is only required to operate for sufficient time to activate the analogue–digital converter at the end of each dwell time. The interruptions in B_2 need then be quite a small proportion of the total time (Fig. 7.12).

We have already seen that it is possible to invert selectively a spin population, and if the INDOR experiment is modified in this way then the perturbation of the irradiated resonance is increased and the sensitivity is increased. The experiment also becomes more compatible with an FT spectrometer since the selective inversion can be done just before the B_1 pulse and there is no possibility of interference. Again, the difference mode is used but with alternative FIDs

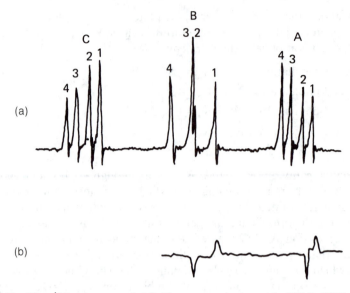

(a)

(b)

Fig. 7.11   The ^{1}H spectrum of 1,2-dibromopropionic acid, $CH_2BrCHBrCOOH$, at 100 MHz in deuteriobenzene with the FT INDOR spectrum below obtained when irradiating the line C^4. (Reproduced with permission from Feeney and Partington (1973) *J. Chem. Soc., Chem. Commun.*, 611.)

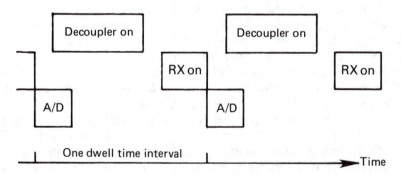

Fig. 7.12   Time-shared homonuclear decoupling. Three parts of the spectrometer have to be switched on and off independently though in a particular order. The receiver alone is switched on for sufficient time to establish an output (RX ON). This output is fed to the analogue-to-digital (A/D) converter, which starts to convert the voltage into a number at a precise time, set by the dwell time in use, and takes a finite time to make the conversion, 5 to 20 µs. The decoupler is switched on during the A/D conversion process and remains on until a little while before the receiver is switched on again. The effect of the decoupler may be modified by reducing the length of each pulse as required.

produced with and without inversion. With inversion, the level 1 population changes from $N + \frac{1}{2}N_a + \frac{1}{2}N_x$ to $N + \frac{1}{2}N_a - \frac{1}{2}N_x$. This is twice the change obtained with saturation, and so the signal intensity is doubled in the difference spectrum. The increase in intensity is even more pronounced in a heteronuclear experiment, where a nucleus with large γ is inverted and one with small γ observed, such as the nuclei ^1H and ^{13}C. If, in Fig. 7.10, we make the A nucleus ^{13}C and the X nucleus ^1H, and invert the X_1 transition, then the populations of levels 1 and 2 are effectively interchanged. The difference in the populations of levels 1 and 3 (A_1) was N_a. Now it is $(N + \frac{1}{2}N_a - \frac{1}{2}N_x) - (N - \frac{1}{2}N_a + \frac{1}{2}N_x)$, which is $N_a - N_x$. The ratio of new signal intensity to old is

$$(N_a - N_x)/N_a = (\gamma_C - \gamma_H)/\gamma_C = -3$$

Similarly, the intensity of the A_2 transition is increased by a factor of $+5$. More complex multiplicities lead to even greater increases in intensity. An example is given in Fig. 7.13 for the carbonyl carbon of acetone. The proton spectrum of acetone contains two sets of ^{13}C satellites, one pair widely spaced and due to coupling to directly bonded ^{13}C in the methyl groups, and one with a much smaller coupling constant due to the two-bond coupling between the methyl protons and the carbonyl ^{13}C. The carbonyl ^{13}C resonance is therefore split into a septet with line intensities 1:6:15:20::15:6:1 by this coupling. If we invert the proton spins corresponding to one of the inner ^{13}C satellites by irradiating this with a long, selective 180° pulse, and follow this immediately with a pulse at the ^{13}C frequency to stimulate a ^{13}C FID, we find that the carbonyl carbon intensities have been greatly perturbed and now have the intensities $+25$: $+102$: $+135$: $+20$: -105: -90: -23, which is a useful gain in intensity. The improvement is particularly marked for the outer lines of the septet, which are not visible in the normal spectrum.

7.6   The INEPT experiment

The title of this section does not mean quite what it appears to do; rather, it illustrates the propensity of NMR spectroscopists to coin acronyms, and INEPT actually stands for insensitive nuclei enhancement by polarization transfer. This is essentially a double-resonance technique, but one in which short pulses are applied to both nuclei with the object of moving the magnetization about in the rotating frame in a known way. It is a good illustration of how pulse sequences can be used to achieve a particular objective.

The low-power methods just discussed are selective in nature, which has the advantage that it is easy to follow what one is doing but the disadvantage that a complex spectrum requires many experiments and much time to understand it fully. The high-power methods are not selective but the spin coupling information is lost. INEPT permits the experimenter to explore the whole spectral region and retain the coupling multiplicities. Evidently, we will be

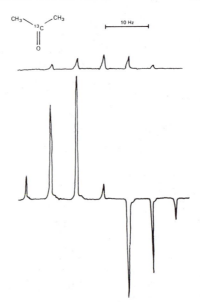

Fig. 7.13   The ^{13}C spectra of the carbonyl group of acetone at natural abundance of ^{13}C. The spectrum is a septet. Two corresponding weak satellites appear in the proton spectrum on either side of the line for the all ^{12}C isotope. If the high-field proton line is inverted using a 180° selective pulse, and then the ^{13}C spectrum stimulated by a 90° read pulse, large enhancements in the intensity of the lines of the septet the obtained as shown in the lower trace. $^{2}J(CH) = 5.92$ Hz. (Reproduced with permission from Jakobsen *et al.* (1974) *J. Magn. Reson.*, **15**, 385.)

interested in the interpretation of ^{13}C spectra, and if multiplet splitting is retained we need also to have a method that increases the intensity of the resonances. INEPT does all this by arranging selective inversion of all the ^{1}H-^{13}C proton doublets at the same time. We proceed as follows, dealing with a single ^{13}C–^{1}H directly bonded pair of atoms.

Initially, the proton magnetization is all in the *z* axis. We define a direction for the *x* axis by applying a 90° pulse to the protons at some phase angle ϕ of the radio frequency, which will swing the magnetization around this axis and into the *xy* plane. The protons coupled to ^{13}C give a doublet, so that the magnetization contains two components, which rotate at different frequencies and so start to separate. This behaviour is illustrated in Fig. 7.14, where the rotating-frame frequency has been chosen so that one component goes faster and one slower, by equal amounts. We now wait a time τ, which is set at one-quarter the reciprocal of the value in hertz of the coupling constant. In this time the two components move apart by an angle of 90°. We then apply a 180° proton pulse with the same phase ϕ, which swings the magnetization around the *x* axis into the opposite side of the *xy* plane. Simultaneously, we invert the ^{13}C magnetization using a 180° carbon pulse. If this was not done, the two proton vectors would converge to form an echo. Inverting the ^{13}C spins, however, interchanges

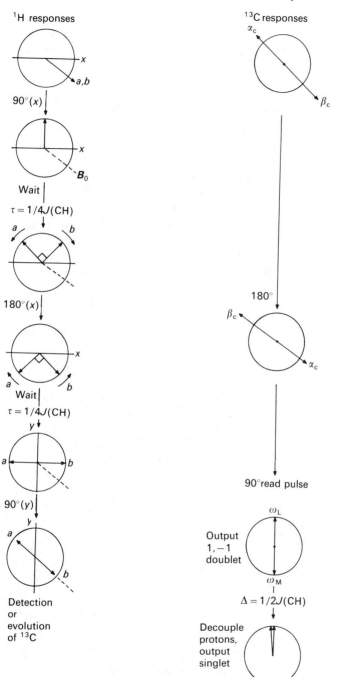

Fig. 7.14   Evolution of magnetization of ^1H and ^{13}C during INEPT experiment. ω_L, ω_M are the two components of the ^{13}C doublet due to coupling to ^1H.

the frequencies of the two proton vectors, so that they diverge, and after a second period of time τ are opposed and lying in the x axis. The phase of the proton pulse is changed by $90°$ to $\phi \pm 90°$ and a $90°$ pulse applied, which now swings the magnetization around the y axis, so selectively inverting one component of the proton doublets. A $90°$ pulse is also applied at the ^{13}C frequency and swings the carbon magnetization into the xy plane, where it produces a detectable output. The inversion of the proton populations provokes changes in the ^{13}C populations also, so that one component of the carbon magnetization is inverted relative to the other and initially is opposed to it. The two components of the magnetization rotate at their individual frequencies and the FID contains two signals corresponding in frequency to the two carbon lines, but with one inverted relative to the other and both enhanced in intensity. The experiment can be further extended to differentiate carbon atoms with one, two or three directly bonded hydrogen atoms. If we allow the carbon magnetization to evolve, the relative positions of the two magnetization components will change. After a time $\Delta = 1/2J(HC)$, they will coincide. If we now apply broad-band proton high-power irradiation, continuously, we remove the proton coupling and equalize the rates of rotation of the two components. If we start the collection of data at this point, then the output is a strong singlet, whereas, if we applied the irradiation at $\Delta = 0$ or $\Delta = 1/J(CH)$, there would be no output since the components would be opposed. On the other hand, if the carbon was part of a CH_2 group and its resonance a 1:2:1 triplet, then the outer lines of this multiplet would cancel for $\Delta = 1/2J(CH)$ and would be reinforced after $\Delta = 1/4J(CH)$. The experiment produces similar output for all CH_n fragments since the values of $^1J(CH)$ are all of a similar order of magnitude (ca. 125 Hz, so τ will be about 2 ms) but can be tuned to be optimum for other values of coupling constant, such as long-distance coupling to quaternary carbon atoms, for instance. This is of particular interest since not only is there an intensity enhancement but the effective relaxation time in the experiment is that of the

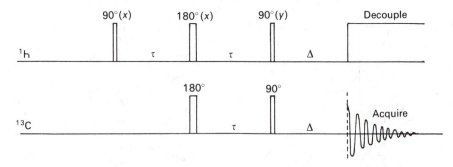

Fig. 7.15   Pulse timing for INEPT experiment shown for both ^1H and ^{13}C frequency channels. The experiment terminates after the $90°$ y pulse and the data are then acquired immediately without proton irradiation. The experiment may be extended by adding an evolution period Δ and then acquiring data with proton irradiation.

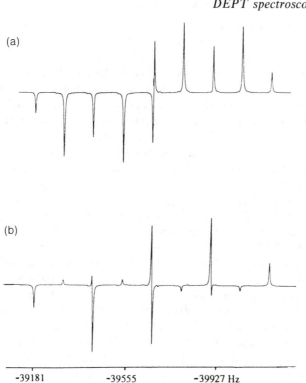

(a)

(b)

-39181   -39555   -39927 Hz

Fig. 7.16   The ^{119}Sn INEPT spectra of EtSn(OAc)$_3$ obtained for different values of τ. (a) $\tau = 0.005$ s corresponding to J(Sn–H) = 50 Hz is longer than the ideal but shows up all nine expected lines. (b) $\tau = 0.0021$ s corresponding to J(Sn–H) = 119 Hz is correctly $1/4J$ for the methylene protons but is double the required length for the methyl protons and so does not invert these and lines are nulled. (From Gouron *et al.* (1990) *Magn. Reson. Chem.*, **28**, 756; copyright (1990) John Wiley and Sons Ltd, reprinted with permission.)

protons and not the usually rather long one of the isolated carbon atoms. Spectral accumulation can thus be faster, giving an even greater reduction in time requirements. The pulse timing diagram is given in Fig. 7.15.

An example of an INEPT spectrum is given in Fig. 7.16 for the ^{119}Sn resonance of ethyltin triacetate, $CH_3CH_2Sn(OAc)_3$. The tin is coupled to both the methyl protons ($^3J(^{119}Sn–^1H) = 242$ Hz) and the methylene protons ($^2J(^{119}Sn–^1H) = 118$ Hz) of the ethyl group. ^3J is almost exactly twice ^2J, and so there is overlap of lines in the triplet of quartets that would be expected and the actual multiplet has nine lines. The form of the INEPT response depends strongly upon the value of τ chosen, because the two coupling constants are different, and the technique can be used to investigate coupling in more complex molecules where the proton spectra may not be very informative because of overlap.

7.7   DEPT spectroscopy

It will be evident from the last example that there are distortions of both phase
and intensity, particularly sign of intensity, in the INEPT experiment. For this
reason, and to produce a more easily handled pulse sequence, a modified technique
has been introduced called 'distortionless enhancement by polarization transfer'
or DEPT. It produces better looking spectra but is unfortunately more difficult
to understand. The pulse timing diagram is shown in Fig. 7.17, and at first
glance seems very little different from the INEPT sequence. We will consider
its effect on a CH_n fragment, where n may have the values 0 to 3. In such a
fragment, the ^1H spectrum will be a doublet due to coupling to ^{13}C and there
will be two proton frequencies. The ^{13}C multiplicity will depend upon n and
there will be from one to four ^{13}C frequencies. The first 90° ^1H pulse swings
the proton magnetization into the xy plane, where the two components precess
at their different frequencies and so separate. The delay 2τ is chosen to equal
$1/2J(CH)$, double that used in the INEPT experiment. At the end of this time
the two proton vectors will be aligned in the x axis. The 90° pulse then applied
at the ^{13}C frequency swings the ^{13}C spins also into the xy plane and the ^1H
nuclei lose their difference in precession frequency: the spin system is static
during the second 2τ period. The 180° pulse at the proton frequency has no
effect except that it refocuses chemical shift effects where several CH_n fragments
are present. However, mutual spin flips occur in the static system and, depending
upon n, a proportion will be mutual $^1H–^{13}C$ flips allowing n to influence the
system response. The third θ pulse to the protons has a value between 45° and
135° and swings the two proton spin vectors into the z direction, or partially
so, and gives an inversion of the magnetization of one of the proton signals,
as required for polarization transfer. The ^{13}C vectors now precess at different
frequencies, and after the third 2τ waiting period will have an orientation in
the xy plane that depends upon n. If we have a CH fragment, then the two ^{13}C
spin vectors will coincide and the output will be a 1:1 doublet, enhanced in
intensity and with neither line inverted as is the case in INEPT. The fully
coupled spectra are obtained if the decoupling is omitted. This can, however,

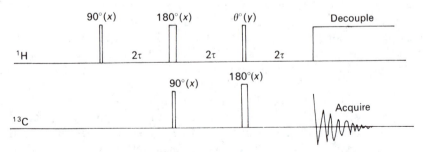

Fig. 7.17   Pulse timing for the DEPT experiment.

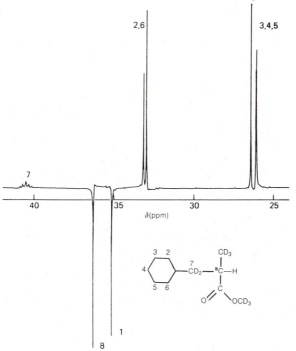

Fig. 7.18   The ^{13}C DEPT spectrum of the monosubstituted cyclohexane shown. (From Ito *et al.* (1989) *Magn. Reson. Chem.*, **27**, 273; copyright (1989) John Wiley and Sons Ltd, reprinted with permission.)

produce congested spectra, and it is better to decouple and use what are called edited spectra to obtain the values of *n*. Since the output depends upon the values of both θ and *n*, it is possible by choosing three values of θ (45°, 90°, 135°) to obtain three spectra, which can be combined to give sub-spectra containing only the resonances of CH or CH_2 or CH_3 carbon atoms. Comparison with the normal ^{13}C spectrum allows the quaternary carbon atom signals to be identified. Alternatively, the value of θ may be chosen to give spectra in which signals from fragments with *n* even have the opposite sign from those with *n* odd. An example of this is shown in Fig. 7.18 for the monosubstituted cyclohexane, $C_6H_{11}CD_2CH(CD_3)(COOCD_3)$. The CH_2 ring carbon atoms give positive signals and the CH carbon atoms in both ring and substituent give negative signals. Some idea of the gain in sensitivity can be obtained from the C7 signals near 40 ppm, which comes from the CD_2 carbon atom and has no polarization transfer effect. The example is also of interest in that one would expect that the carbon atoms 2 and 6 would have the same chemical shift and equally the carbon atoms 3 and 5. This is patently not the case, since five ring carbon signals are resolved, and is believed to arise because of the asymmetry at C8.

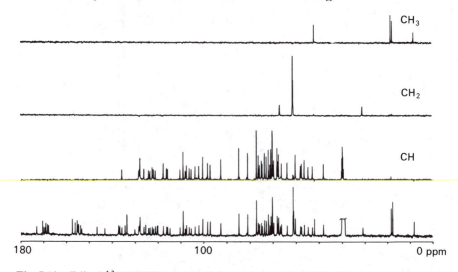

Fig. 7.19   Edited ^{13}C DEPT spectra of ristocetin at 100.6 MHz. The lower spectrum is the normal broad-band decoupled spectrum showing all the carbon resonances. Note that even the resonance of the CD_2H groups of the solvent dimethylsulphoxide-d_6 appears in its appropriate trace.

This example is relatively simple, though illustrative. The method is immensely useful in the case of complex molecules. An example is given in Fig. 7.19 for the antibiotic ristocetin, which has a molecular of weight of about 2000.

7.8   Proton homonuclear double resonance

This has to be a low-power technique, since the idea is to perturb one multiplet in a spectrum and observe what happens elsewhere in the spectrum. Selective irradiation is used at a power sufficient for a local perturbation but not so high as to be detected in the rest of the spectral region.

Prior to the days of 2D NMR, such experiments were carried out to discover the connectivities between proton resonances via the spin–spin coupling interaction. It was particularly useful where there was overlap of multiplets since the changes in spectral envelope were usually quite evident. However, the more complex the molecules studied, and so the more complex their spectra, the greater the number of such experiments that had to be carried out and the more time-consuming these became. A 2D spectrum would now be obtained to carry out the same task and, currently, molecules as complex as proteins are investigated in this way. An example of the 1D technique, as the experiment would now be called, is given for the spectrum of the palladium complex shown in Fig. 7.20. The whole proton spectrum is shown at the foot of the figure, with

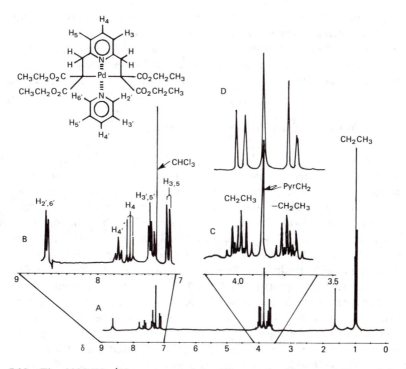

Fig. 7.20 The 200 MHz ^{1}H spectrum of the Pd complex illustrated. Trace A is the normal spectrum, with the line at 0 ppm being the reference TMS. The solvent was CDCl$_3$, though this contains a trace of the protonated solvent, which shows as a singlet at 7.2 ppm. Traces B and C are expanded versions of A. Trace D was obtained while irradiating the triplet at 1 ppm. (From Newkome *et al.* (1981) *J. Am. Chem. Soc.*, **103**, 3425; copyright (1981) American Chemical Society, reprinted with permission.)

two portions shown expanded above this. The lines are assigned by lettering or underlining to all the protons in the complex. All the resonances are well separated and the assignment is straightforward A$_2$X for the substituted pyridine ligand, [M]$_2$A$_2$X for the pyridine ligand, a singlet for the methylene group attached to the one pyridine ligand, a triplet for the methyl group of the carboxyethyl substituents and, by difference, a complex, 16-line multiplet for the corresponding methylene protons. The chemical shifts and integral trace (not shown) all confirm these assignments. The question arises of why the ethyl methylene signals are so complex. The singlet of the methylene groups directly bonded to pyridine indicates that the molecule has a plane of symmetry, and we would expect then that the CO methylene protons were equivalent also and would give a 1:3:3:1 quartet. The effect of irradiating the methyl triplet at 1 ppm was therefore studied, sufficient power being used to eliminate the splitting due to coupling completely. The only changes observed are demonstrated in trace D, where the methylene spectrum has been simplified to an AB quartet with a large geminal copling constant. The two protons on each methylene group thus must have different chemical shifts. The free rotation of the ethyl groups thus does not in this case result in an averaging of the methylene proton environments despite the symmetry of the molecule as a whole, since there is no plane of symmetry through the bonds attaching the methyl groups to the complex.

7.9   Selective nuclear Overhauser enhancements

So far we have only discussed the NOE in the presence of strong, broad-band irradiation or the generalized NOE that occurs if one transition is weakly irradiated. If the whole resonance of a proton is irradiated, whether it be a singlet or a multiplet due to spin–spin coupling, this will cause intensity changes to the resonances of other nuclei that are spatially close. The enhancements possible are quite small: $1 + \frac{1}{2}\gamma_H/\gamma_H$ means that $\eta = 1/2$ only. It is usual then to obtain such spectra by using the difference mode, i.e. spectra are obtained with and without double irradiation, the FIDs subtracted and the difference Fourier-transformed to give a spectrum that ideally contains responses only from those resonances perturbed in intensity by the double irradiation. An example is shown in Fig. 7.21 of a cobalt complex that carries four equatorial methyl groups situated in the the CoN_4 plane and a pendant n-butyl group. Is it possible to determine how this latter is disposed relative to the plane of the complex? The 400 MHz proton spectrum of the butyl part of the complex is shown in the lower trace and the NOE difference spectrum obtained while irradiating the methyl group signal is shown above the spectrum. It is clear from the interal traces that two of the butyl multiplets show an NOE and two do not. The obvious disturbance at the position of the triplet, which must be due to a small frequency shift, integrates to zero. If we can assign the individual resonances, then we will know which are nearest to the equatorial methyl groups. Thus the triplet to high field is obviously the δ-CH_3. The sextet at 1.23 ppm is the γ-methylene, which is coupled by roughly equal amounts to five protons; similarly, the quintet is due to the β-methylene. The α-methylene also gives a triplet, though distorted due to second-order effects, which probably indicate that there is some restriction of rotation at this end of the substituent. We then see that it is only the $\alpha + \beta$ groups that show an NOE. Since relaxation phenomena and NOE fall off as r^{-6}, this indicates that the other two groups are even further away from the equatorial methyl groups and that the butyl substituent must be normal to the plane of the CoN_4 ring and not bent as shown in the formula. A similar NOE is also observed for the protons 2 in the pyridine ligand (not shown).

A more complex example is that of deciding the conformations in solution of various erythromycin A derivatives. The formula of erythromycin A is shown in Fig. 7.22, together with two crystal structures that have been determined for different derivatives and labelled A and B. These structures vary as a function

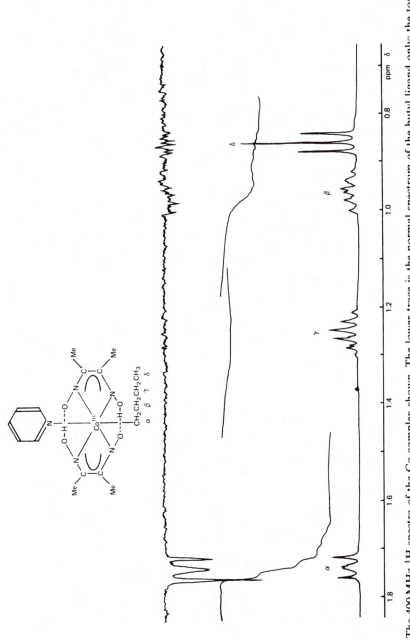

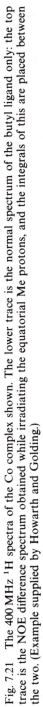

Fig. 7.21 The 400 MHz ^{1}H spectra of the Co complex shown. The lower trace is the normal spectrum of the butyl ligand only; the top trace is the NOE difference spectrum obtained while irradiating the equatorial Me protons, and the integrals of this are placed between the two. (Example supplied by Howarth and Golding.)

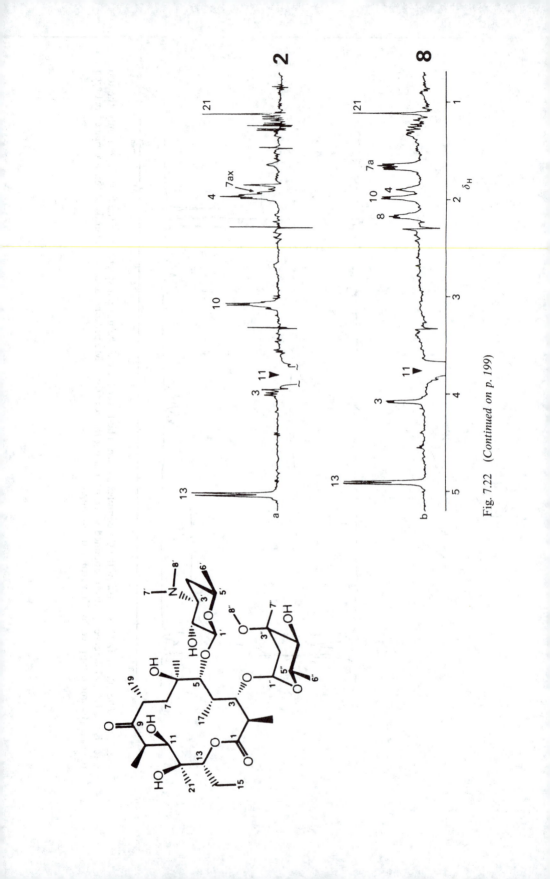

Fig. 7.22 (*Continued on p. 199*)

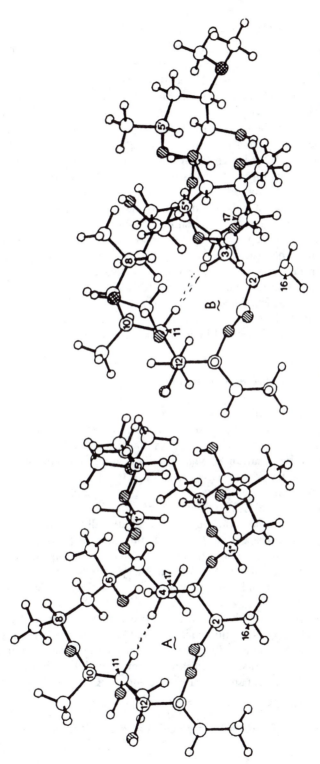

Fig. 7.22 Formula structures and NOE difference spectra (400 MHz, ^{1}H) of erythromycin A and its 9-hydroxo derivative. (From Everett *et al.* (1990) *Magn. Reson. Chem.*, **28**, 114–18; copyright (1990) John Wiley and Sons Ltd, reprinted with permission.)

of different substituents at the 9 and 11 positions of the 14-membered lactone ring of the parent substance. The 400 MHz NOE difference spectra are shown of erythromycin A (upper trace 2) and the derivative where the 9-keto group has been replaced by a 9-hydroxo substituent (lower trace 8). Irradiation was applied to the resonance of proton 11, and enhancements were observed to protons close by on the same portion of the ring (10, 21, 13) and to protons that are further away along the ring but brought near by the conformation of the ring. Compound 2 exists in the A conformation since it has a large NOE to proton 4, which is turned towards proton 11, and small NOE to proton 3, which is turned away. In compound 8 the NOEs to both protons 3 and 4 are comparable, and here it is believed that we have conformation B with some exchange with conformation A. It should be noted that, as in the previous example, a necessary step in the interpretation of such spectra is knowing the full assignment of the proton spectrum. In the case of a complex molecule such as that illustrated, this is not a trivial problem, and involved a series of experiments: resolution-enhanced ^{1}H spectra, DEPT spectra, spin-echo and normal broad-band decoupled ^{13}C, and three types of 2D experiment were necessary in addition to establish the various connectivities between protons. Examples of the techniques will be found in the next chapter.

Exercises in spectral interpretation

The determination of the structures of organic molecules using proton and carbon spectra together will now be considered in some exercises.

We have shown in the previous chapters that the carbon spectra are capable of yielding much information through a series of quite time-consuming experiments. These facilities are normally reserved for the more difficult samples, and often it will be sufficient to have simply the broad-band proton-decoupled spectrum. This gives the chemical shift information, details of coupling to nuclei other than hydrogen, and a carbon count, provided all the likely errors in line intensity are taken into account. It is also certain that the proton spectrum will be available, since this is so easy to obtain, and interpretation will be based on the two sets of data taken together. The carbon spectra, of course, give information about atoms not bonded to hydrogen, which is not available in the proton spectra.

The carbon chemical shifts are much more widely dispersed than are the proton shifts. The ranges within which different types of carbon atom resonate are shown on the chart in Fig. 7.23. The chemical shift of a given carbon atom in a family of compounds is sensitive to the influence of all four substituents, and for alkanes, for instance, can be predicted using the Grant and Paul rules:

$$\delta_i = -2.6 + 9.1 n_\alpha + 9\ 4 n_\beta - 2.5 n_\gamma + 0.3 n_\delta$$

The chemical shift of carbon i can be calculated from the number of directly

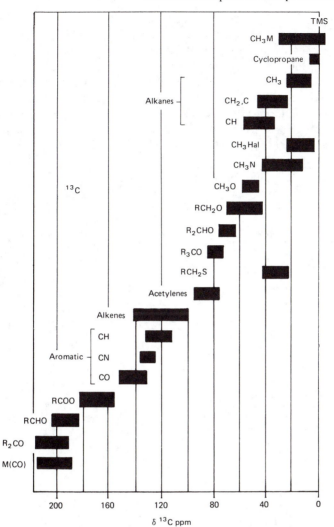

Fig. 7.23 Chart of approximate chemical shift ranges of carbon atom nuclei in organic and organometallic compounds. M represents a metal and Hal a halogen.

bonded carbons (n_α) and the number of carbon atoms two (n_β), three (n_γ) and four (n_δ) bonds removed; -2.6 ppm is the chemical shift of methane. Similar rules exist for ethylenic hydrocarbons and substituent effects have been documented.

The use of the two sets of information together is illustrated in exercise 13 for the simple case of ethylbenzene. The 60 MHz proton spectrum contains the unmistakable quartet–triplet pattern due to an ethyl group and five protons resonate in the aromatic region. This is indeed sufficient to give the structure in this case, especially if the molecular weight were known. However, we note for the purposes of illustration that there is no structure in the aromatic resonance and we would find that no useful improvement could be obtained even at the highest fields. We therefore turn our attention to the carbon spectrum, also in exercise 13. We find the two carbons of the ethyl group and four aromatic resonances. The one to low field is rather small, as would be expected for quaternary carbon. Of the remaining three, two are significantly larger and might arise from two carbons. The pattern has the form to be expected for a monosubstituted six-membered ring, and allows us to overdetermine the structure of this molecule. All its features are apparent, though we should note carefully the ambiguities of the intensity of the resonances. Which line should we take as representing one carbon? Some more complex examples are given in exercises 14–16.

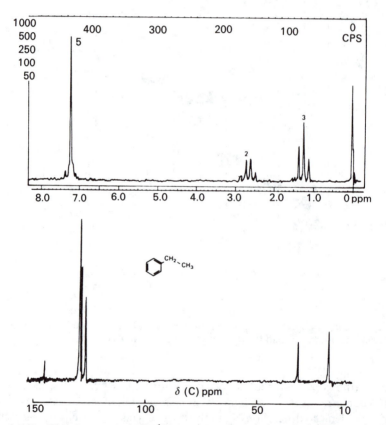

Exercise 13   The upper trace is the ^{1}H spectrum of a solution of ethylbenzene in CDCl$_3$ obtained at 60 MHz and with TMS added as reference. The lower trace is the ^{13}C spectrum obtained at 22.6 MHz with proton broad-band decoupling at 90 HMz. No TMS was present in this case, the calibration being based on the deuterium lock frequency. (Upper trace reproduced by permission of Varian International AG.)

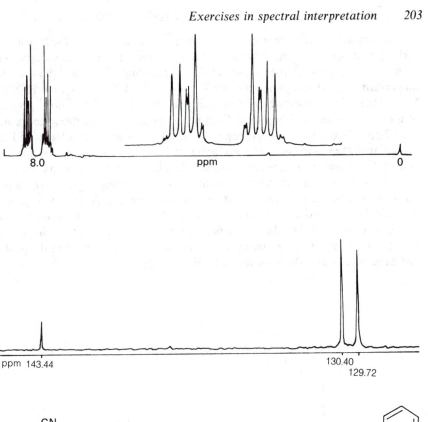

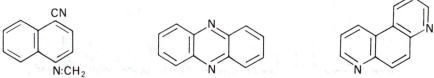

Exercise 14   The 80 MHz proton spectrum (upper) and 20.1 MHz ^{13}C spectrum (lower) of the aromatic compound $C_{12}N_2H_8$. The proton spectrum is diagnostic of an $[AB]_2$ four-spin system in which there are two pairs of protons with the same chemical shift δ_A, δ_B, but different coupling constants $J(AB)$, $J(AA')$, $J(AB')$ and $J(BB')$. Only three lines are observed in the carbon spectrum. This information is sufficient to identify the compound, some suggestions for which appear below, the spectra. (Example provided by A. Römer.)

Exercise 15(a) shows two carbon spectra of a compound related to that used to obtain exercise 14, but synthesized so as to contain the substituents CH_3-, CH_3O- and CH_3OCO- on the aromatic core. The empirical formula is $C_{16}H_{14}O_3N_2$ and this and the parent compound have very similar UV spectra, with a maximum at 360 nm. The resonances are displayed in two groups, 0 to 70 ppm and 95 to 170 ppm. In each group the lower spectrum is with broad-band proton irradiation and the upper, noisy, one is without and so displays all the carbon–hydrogen couplings, whose values are included on the spectra. The large values represent the one-bond C–N coupling. Typical values of the longer-range C–H couplings are HCC $^2J \sim 1$, HCCC $^3J \sim 9$ and HCCCC $^4J \sim 1$ Hz in these aromatic systems. Exercise 15(b) shows the proton spectrum of the same molecule. The broad lines near 1 ppm are impurities, the quintet near 2 ppm is the remnant hydrogen in the deuterioacetone solvent and the singlet at 2.7 ppm is water. The multiplet at 2.8 ppm is a doublet of doublets and that at 4 ppm is an unequal doublet. The lower-field half of the multiplet near 7.3 ppm (due to two protons) contains one coupling equal to the larger one observed in the multiplet at 2.8 ppm, and the broadening in the high-field half can be assigned to the other coupling, though this is unresolved in the aromatic resonance. The data in the three spectra are sufficient to allow a full structural assignment to be made, taking into account the pattern in the proton aromatic region and the long-range C–H couplings.

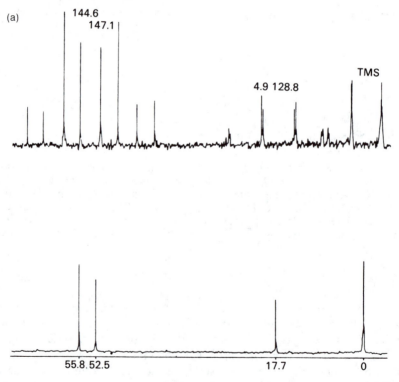

(a) 144.6 147.1   4.9 128.8   TMS

55.8 52.5   17.7   0

(continued)

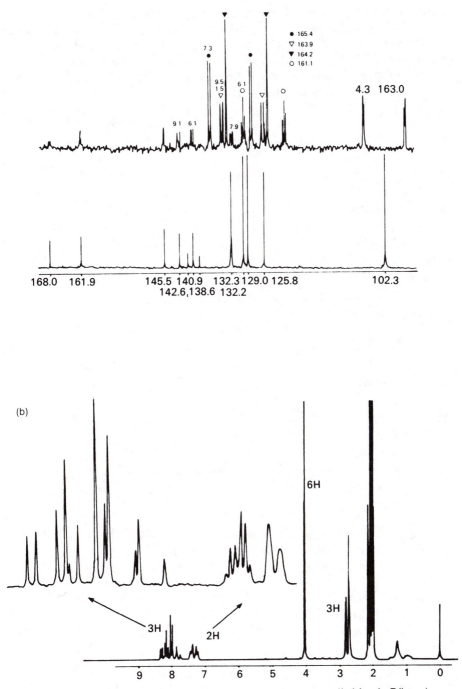

Exercise 15   (a) ^{13}C spectra; (b) ^{1}H spectra. (Example supplied by A. Römer.)

Exercise 16 shows two ^{13}C spectra of the compound $C_{10}H_2O_6$, one with (b) and one without (a) proton double irradiation. Coupling to hydrogen is observed in the latter, $J = 179\,Hz$ to the resonance at 122 ppm and $J = 6\,Hz$ to the resonance at 138 ppm, though this cannot be seen in the trace illustrated. The proton spectrum is a singlet in the aromatic region. What is the structure of the molecule, given that it is an anhydride?

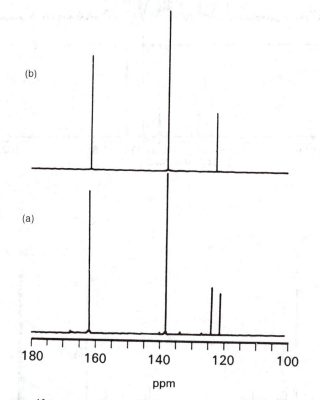

Exercise 16   Two ^{13}C spectra. (From Wolfe *et al.* (1991) *Magn. Reson. Chem.*, **25**, 441; copyright (1991) John Wiley and Sons, reprinted with permission.)

8 Two-dimensional NMR spectroscopy

Two-dimensional NMR spectroscopy, usually referred to as 2D NMR, has become an important tool for chemists. Several experiments are possible and new variants are being continually invented, such is the power of the method. The two dimensions are dimensions of time. One of these is already familiar to the student, and is the time domain with which we collect the FID output from the spectrometer and which contains frequency and intensity information. The second dimension refers to the time elapsing between the application of some perturbation to the system and the onset of the collection of data in the first time domain. This second time period is varied in a regular way and a series of FID responses collected corresponding to each period chosen. Each FID is Fourier-transformed to produce perturbed spectra, which are then stacked, and the sets of data at each point of frequency transformed again (one can think in terms of two transforms of the data taking place at right angles) to give data that can be displayed on two frequency axes (e.g. ^1H chemical shift versus $^1H-^1H$ couplings, or ^1H chemical shift versus ^{13}C chemical shift of spin–coupled pairs), with some means of indicating intensity – a stacked plot of absorption spectra as seen previously, or each peak is represented by contours as in a relief map. The plot may or may not contain the data present in the 1D representation but will, in addition, contain correlations between various parts of the total spin system examined, which will permit the connectivities between different nuclei such as spin–spin coupling or spatial proximity (i.e. NOE) to be established and so allow the spectra to be assigned to the constituent atoms of the molecule. The nature of the method is represented in Fig. 8.1. We will describe some of the experiments possible in more detail below. If the molecule to be studied is complex, it is necessary to carry out several different 2D experiments in order to discover all the correlations necessary to provide an unequivocal structure. We will deal with the elucidation of the structure of amygdalin, whose formula is in Fig. 8.2, though we will describe each technique in succession and then refer back to the problem of amygdalin.

8.1 *J*-resolved 2D spectroscopy

This technique is based on the Carr–Purcell pulse sequence described in section 5.13.2 for the determination of T_2, which only works if the resonances studied

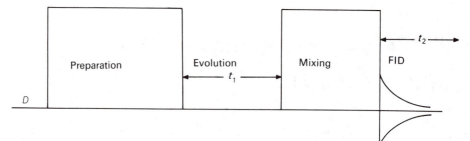

Fig. 8.1   The component parts of a 2D experiment. *D* is a pre-experimental delay to allow the system to come to equilibrium. A variety of pulse sequences may be applied during the mixing and preparation boxes. t_1 and t_2 are the two time dimensions, which appear as variables in the final spectra, though as frequency dimensions after Fourier transformation.

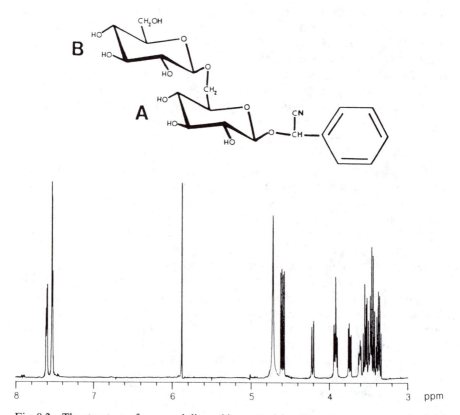

Fig. 8.2   The structure of D-amygdalin and its normal, i.e. 1D, proton spectrum obtained at 500 MHz. There is a pattern in 2:3 intensity ratio at 7.6 ppm due to phenyl, a CHCN singlet at 5.88 ppm, HOD at about 4.75 ppm and a complex pattern to high field due to the remaining 14 sugar protons. (From Ribiero (1990) *Magn. Reson. Chem.*, **28**, 765–73; copyright (1990) John Wiley and Sons Ltd, reprinted with permission.)

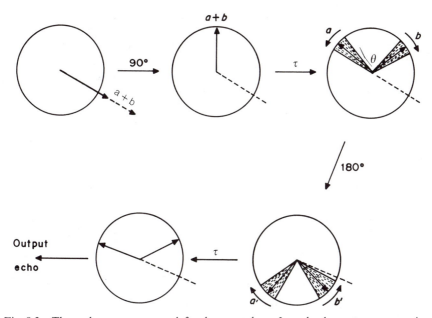

Fig. 8.3   The pulse sequence used for homonuclear *J*-resolved spectroscopy. τ is a waiting time, $2\tau = t_1$.

are singlets, since the refocusing of doublets caused by spin–spin coupling modulates the output intensity as a function of the waiting time. This is easily transformed to advantage into a 2D experiment. The pulse sequence used and the behaviour of the magnetization are shown in Fig. 8.3. We will consider the behaviour of the magnetization of a doublet in isolation, the doublet splitting being caused by another spin-1/2 nucleus of the same type. Initially, the magnetization is in the direction of $\boldsymbol{B}_0$ with the two components a and b precessing at different velocities around the field direction. The magnetization is swung into the xy plane by a 90° pulse, and the different precession frequencies now cause the a and b components to separate by an angle θ after some time τ. In addition, these magnetization vectors also start to fan out, as shown by the shaded areas, because of the magnetic field inhomogeneities, which are the main cause of line broadening in many high-resolution proton spectra. At the end of time τ, the sample is subjected to a 180° pulse, which is shown as inverting the magnetization vectors around the x axis. However, the same pulse also inverts the magnetization of the coupled nucleus, with the result that the rates of rotation of the a and b vectors are interchanged; a' now rotates as old b and b' as old a. The magnetization vectors continue to separate but the inhomogeneity loss of coherence is still refocused, as in the Carr–Purcell experiment. Thus, after another time period of τ, the inhomogeneity broadening has gone but the two vectors are at some angle whose value depends upon the value chosen for τ. The echo thus produces the usual FID, which contains the

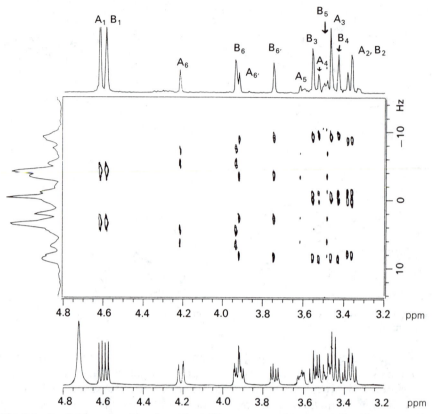

Fig. 8.4   Partial homonuclear *J*-resolved 2D spectrum of D-amygdalin. The responses are shown as contour plots on the *J*/δ chart. The normal 1D spectrum is below and the fully decoupled version of this is above the chart. All coupling constants can be read off on the right-hand scale. (From Ribiero (1990) *Magn. Reson. Chem.*, **28**, 765–73; copyright (1990) John Wiley and Sons Ltd, reprinted with permission.)

frequencies of the two lines of the doublet but which have initial phase angles that differ from those that both would have immediately after a 90° pulse. The output thus contains both frequency and phase information. By repeating the experiment for various values of τ, a series of FID responses is obtained, which can be transformed in two directions and the resulting data manipulated to give a contour plot in which the chemical shift of the nucleus studied appears along one axis and the coupling constants along the other. The technique thus effectively turns the multiplet structures round 90° so that they no longer appear in the chemical shift dimension and overlap is avoided.

A typical plot is shown in Fig. 8.4 for the high-field sugar resonances of amygdalin, where most spectral congestion lies. The chemical shifts are plotted on the horizontal scale, whose length is equivalent to 800 Hz, and coupling constants are in the vertical axis. The plot below the chart is the normal 1D spectrum. That at the top is the projection of the contour plots onto the chemical shift axis and is, in fact, a spectrum in which all the protons have been fully decoupled from one another, i.e. the experimentally unachievable homonuclear broad-band decoupling experiment. Clearly, this plot helps to distinguish resonances that overlap in the 1D trace. There are, for instance, two 1:1:1:1 quartets at 4.6 ppm and 3.73 ppm, and we can see immediately that the former consists of two equal overlapping doublets whereas the latter is a doublet of doublets involving only one chemical shift value. The multiplet structure appears in the coupling dimension. 128 values of τ were used and 128 FIDs were obtained, which limits the digital resolution to 0.39 Hz. It is a simple matter to obtain all the coupling constants to within this limit. What can we deduce about the spectrum of D-amygdalin from this chart? The lines are already assigned on the figure but we will attempt to proceed as if this had not already been done. Four lines are marked as originating from protons-6. They are all doublets of doublets, but one of the couplings is rather large at about 11 Hz and is found nowhere else in the spectrum, and we can think immediately in terms of a possible geminal pair. The two large couplings are slightly different, sufficiently so that it is possible to separate the four resonances into two pairs linked by identical coupling constants. These are then the CH_2 protons linked to the two rings, although we do not actually know yet which is which. Each proton of these geminal pairs is further coupled to proton 5, though the responses from both of these is rather weak and we cannot reliably proceed to further assignment. The two doublets near 4.6 ppm have a chemical shift that is characteristic of the anomeric proton H1 of β-pyranosides of glucose. Further, these protons are the only ones that will be split into simple doublets, and so the assignment is confirmed. Finally, we can see that all the coupling constants other than those between protons-6 are about 7–8 Hz, so that we have axial–axial stereo-chemistry rather than equatorial–equatorial or equatorial–axial.

8.2   COSY spectroscopy

COSY stands for correlation spectroscopy, and was the first 2D technique to be proposed. As we shall see, it has given rise to many variants, which depend upon the existence of spin–spin coupling between nuclei to provide supplementary responses that relate the chemical shift positions of the coupled nuclei. It is equivalent to carrying out simultaneously a series of double-resonance experiments at each multiplet in the spectrum and looking for the part of the spectrum where a perturbation has occurred. It is thus rapid and avoids the need to know what irradiation frequency to use for each experiment, information that in any case is often difficult to obtain for a complex or crowded spectrum. Figure 8.5 shows the pulse sequence used. The first 90° pulse flips the *z*

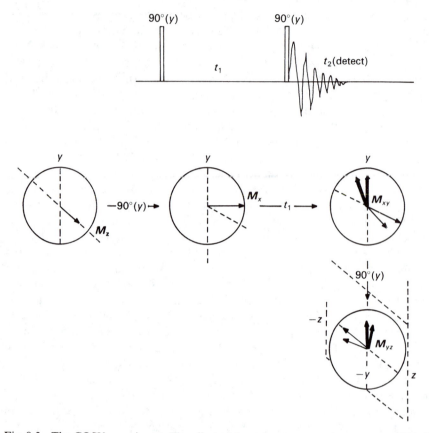

Fig. 8.5   The COSY experiment. The effect of the pulse sequence shown on an AX spin system. The A and X spins are differentiated by the light and heavy lines. The vectors separate slowly under evolution with J and more rapidly under evolution with $v_0\delta$. In the fourth diagram, all the magnetization is found instantaneously in the yz plane.

magnetization into the x direction. If we consider that we have an AX spin system with two doublets due to spin–spin coupling, then the magnetization comprises four components precessing at different frequencies. Thus it separates into two pairs of vectors due to the chemical shift difference, with the separation in each pair being due to the coupling interaction. The exact disposition of all four components will depend upon the value of t_1. The second 90° pulse swings the magnetization instantaneously into the yz plane, where it starts to precess around the z axis and give an output. The position of the magnetization in the yz plane depends upon where the component vectors were placed at the end of t_1. If they were in the right half of the xy plane, then they are swung into the $-z$ direction and are inverted, otherwise they take the $+z$ direction; they will, of course, in general, have some xy magnetization also. Inversion of some of the components causes transfer of polarization, and so the signal of either of the doublets will be a function of t_1 and the chemical shift difference frequency, and will in fact be modulated by the chemical shift and the way this evolves during t_1. The output for each value of t_1 will contain the four frequencies of the two AX doublets, but these will be distorted in both intensity and phase. Thus we obtain several outputs for a range of t_1 values, we Fourier transform the FIDs to obtain spectra with different degrees of distortion and then transform the stack at each frequency position to obtain a 2D map. Where there has been no correlation between spins, then the frequency is the same in both dimensions and the signals appear on a diagonal plot, which is simply the normal 1D spectrum. Where there is a correlation, then the chemical shift and coupling constant are mixed in each resonance and a signal appears off the diagonal at the two points where the coordinates are those of the A and X chemical shifts. We can thus tell instantly where the two coupled spins are situated in the normal 1D spectrum.

A partial COSY spectrum of amygdalin is shown in Fig. 8.6, concentrating on the high-field part of the spectrum where all the interest lies. The diagonal runs from top left to bottom right of the map and when projected onto a chemical shift axis is seen to be the 1D spectrum, in this case shown at the top of the figure. Several off-diagonal responses are evident, disposed symmetrically about the diagonal, and it is these we must examine for information about correlation. The anomeric signals at about 4.6 ppm can be used as a point of entry for analysis of the spectrum, and the connectivities to the H2 signals can easily be traced. Unfortunately, the diagonal is very cluttered in the 3.4 to 3.6 ppm region, and we cannot move on with any degree of certainty to the connectivities to protons further around the ring. This is a disadvantage of COSY, which we will see how to circumvent shortly. The doublet of doublets at 4.22 ppm has a strong cross-peak to the complex multiplet at 3.92 ppm and a weak one to the multiplet at 3.62 ppm, the latter also being correlated with the 3.92 ppm resonance. This pattern is that of an ABX spin system and so has to be due to H6, H6′ and H5 of one ring. Note that the response marked with the asterisk is not directly on the tie line used to trace these connectivities and so is not part of this group. The multiplet at 3.62 ppm is an octet and so has to be H5, in agreement with the *J*-resolved data, which has already allowed us to locate the protons 6. It follows from the cross-peak near the diagonal that H4 of this ring is at about 3.5 ppm. A second, similar pattern, shown by dashed tie lines on the other side of the diagonal, connects signals at 3.9, 3.75 and 3.49 ppm and presumably arises from the protons 5 and 6 of the second ring. H4 cannot be assigned in this case. It remains to assign this and the H3 protons.

Two further points need to be made about this simple COSY spectroscopy. In the first place, while it is not evident in the plot shown, the spin coupling information is also present in the COSY map. Close examination of a cross-peak will show that it is composed of a grid of peaks separated by the coupling constants involved. These peaks have different signs and are usually represented

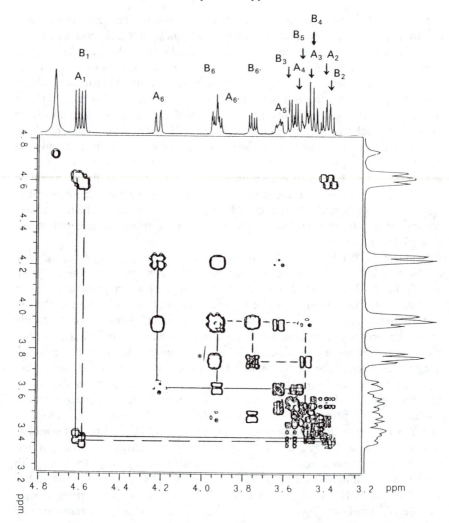

Fig. 8.6   Partial COSY spectrum of D-amygdalin at 30°C showing the sugar resonances. The tie lines relate resonances on the diagonal via the off-diagonal cross-peaks. (From Ribiero (1990) *Magn. Reson. Chem.*, **18**, 765–73; copyright (1990) John Wiley and Sons Ltd, reprinted with permission.)

by different coloured inks in the COSY map. In the case of complex spectra, the *J*-resolved plots become too crowded to be usable and the COSY spectra are then used to extract coupling constants, though with modifications to the technique known as DISCO (differences and sums in COSY) or E-COSY (exclusive COSY), which simplify the patterns in the cross-peaks. The second important point is that the phases of the pulses are often cycled in a predetermined way. Thus, while it is not strictly necessary for a COSY spectrum, it is advantageous if the first pulse sequence is $90°(y)-t_1-90°(y)-t_2$ as discussed, and the next is $90°(y)-t_1-90°(-y)-t_2$, i.e. the phase of the second pulse is changed by $180°$. The two outputs are added and the pair of sequences continued as long as is necessary. The more complex pulse sequences that we will encounter below permit more complex permutation of pulse phases, and quite extended cycles can be used to select a particular type of polarization step for examination and to exclude all others. Pulse cycling is thus an integral part of 2D spectroscopy, without which it cannot operate properly. We will not discuss the phase cycles further here, since this would extend this chapter too much, and we are primarily interested in introducing the possibilities opened up by 2D spectroscopy, rather than indicating how experiments should be set up in detail.

8.3   Zero- and double-quantum COSY

The diagonal trace obtained in a COSY spectrum, while helping to simplify interpretation, also can hide useful information because of its breadth. A way around this is to obtain correlations via polarization transfer involving zero-quantum (ZQ) or double-quantum (DQ) transitions. In a two-spin system, such transitions correspond to the normally forbidden $\alpha\beta \to \beta\alpha$ or $\alpha\alpha \to \beta\beta$ mutual spin flips, respectively. These are not observable since they do not correspond to the $\Delta m = \pm 1$ rule, but their effects on the normal single-quantum transitions may be evident. In spectra run so as to show up the ZQ or DQ polarizations, we obtain the 1D chemical shift-coupling constants on one axis and the algebraic difference between the frequencies (ZQ) or their sum (DQ) on the other axis. There is thus no diagonal to obscure the cross-peaks, though the interpretation of the spectra is less straightforward. We will discuss ZQCOSY as applied to the amygdalin problem, but first we must indicate that DQCOSY is probably more used and that it has a particularly useful place in ^{13}C spectroscopy. In a molecule with ^{13}C at its natural abundance (1.1%), most molecules will contain only one ^{13}C atom, but some will contain two, and in some of these they will be directly bonded. The technique thus probes carbon–carbon connectivities in molecules and is called INADEQUATE spectroscopy because of its extreme usefulness (INADEQUATE = incredible natural abundance double-quantum transfer experiment). Both ZQ and DQ spectroscopy start with a pulse sandwich, which is tuned to the value of coupling constant between the nuclei it is wished to examine. The polarization transfers made possible by this sequence are then allowed to evolve for t_1 and are then converted to observable transverse magnetization by a final $90°$ pulse. The phase cycling of the pulses is important, as it is this which selects the transfer desired. The pulse sequence used for ZQCOSY is shown in Fig. 8.7 and the resulting spectra for amygdalin are in Fig. 8.8. The horizontal axis is the chemical shift axis and the cross-peaks for vicinally coupled protons AX appear at coordinates $(\delta_A, +\Delta v)$ and $(\delta_B, -\Delta v)$. For example, for the H1, H2 protons, whose chemical shifts we have already obtained from the

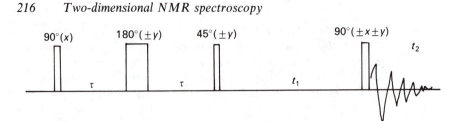

Fig. 8.7 The pulse sequence used for ZQCOSY. In the case of DQCOSY or INADEQUATE, the third pulse is a 90° pulse. In the spectrum shown in Fig. 8.8, the value of τ was 35.7 ms, optimized for a vicinal coupling constant of 7 Hz (τ = 1/4J).

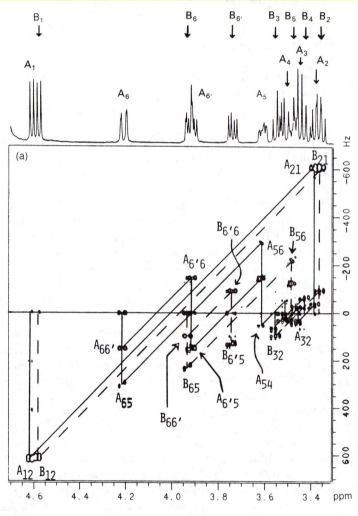

Fig. 8.8   (*Continued on p. 217*)

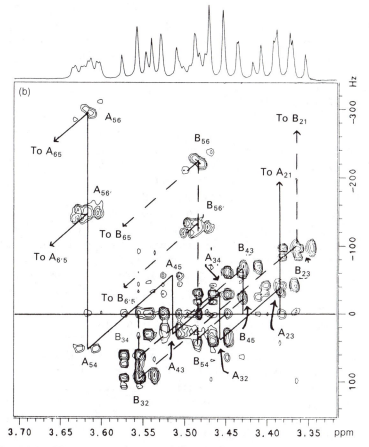

Fig. 8.8   ZQCOSY spectrum of D-amygdalin showing (a) the whole of the sugar region and (b) an expansion of the high-field, crowded part of this region. (From Ribiero (1990) *Magn. Reson. Chem.*, **28**, 765–73; copyright (1990) John Wiley and Sons Ltd, reprinted with permission.)

COSY spectrum, we know the frequency difference to be 500×1.2 ppm or 600 Hz. There are thus H1 peaks at 4.6, 600 and H2 peaks at 3.4, -600. The resonances are marked, e.g. A_{12}, which means the peak due to H1 in ring A, coupled to H2 in ring A. They are connected by a tie line with a slope of 2. If a given proton is coupled to more than one other type of proton, then it has several responses, which all appear at its chemical shift position and so lie on a vertical line in the plots of Fig. 8.8. Thus the A6 proton is coupled both to A6′ and to A5 and shows two responses. These are then related to the 6′ and 5 chemical shifts by lines of slope 2, which meet the appropriate responses an equal distance on the other side of the 0 Hz line. We can then assign the remaining resonances. By dropping a perpendicular from A_{21} we locate A_{23} and a line of slope 2 then finds A_{32} (Fig. 8.8(b)). A perpendicular finds A_{34} and a line of slope 2 A_{43}. Another vertical finds A_{45} and so we proceed to the already assigned part of the spectrum, thus being able to complete in effect a walk around the ring A. Ring B is assigned in the same way. We have thus assigned all the proton resonances of the sugar rings of D-amygdalin.

8.4   NOE and NOESY

It remains to ascertain which of rings A and B of D-amygdalin is attached directly to the CHCN group. This is done simply in this case by irradiating the CHCN proton at 5.88 ppm and observing which of the doublets around 4.6 ppm (the H1 protons) showed a NOE enhancement. It proved to be the one slightly to low field at 4.61 ppm, and this is then in the A ring, as, conveniently, we have labelled the peaks throughout. The inverse experiment is also possible, in which the doublets at 4.61 and 4.58 ppm are irradiated in turn and the effect observed on the CHCN proton. This leads to the same conclusion, though with less clarity, since it is not possible to irradiate such close resonances with sufficient selectivity so that one gives an NOE and the other none. A weaker effect is detected in the latter case due to spill-over of the irradiation power into the 4.61 ppm doublet region.

It would, of course, have been possible to carry out a 2D version of this experiment on D-amygdalin, but since the molecule contains only one interaction of interest the experimental time necessary cannot be justified. Such experiments are, however, very important in the case of large, flexible biomolecules such as peptides. In solution, it is possible, once the proton resonances of identifiable residues have been assigned, to determine which are in close proximity in space. Thus the way the chains of such molecules are folded can be ascertained, and the data currently being obtained in solution studies of large molecules are comparable in accuracy with crystallographic data on the same molecules. This experiment is called Nuclear Overhauser Effect Spectroscopy (NOESY), and while it will not be illustrated in detail here, we will shortly discuss the investigation of chemical exchange by 2D spectroscopy, which uses essentially the same technique.

8.5   Heteronuclear COSY

The next stage in understanding the spectroscopic properties of our model molecule is to assign the ^{13}C resonances. This is done by investigating the connectivities imparted by $^{1}H-^{13}C$ coupling paths. The task is facilitated by the fact that the one-bond CH coupling constants are much larger than the two- or three-bond (CCH or CCCH) coupling constants, so that the pulse sequence used can differentiate between different types of coupling path. Decoupling of protons from ^{13}C is achieved in the usual way by irradiating

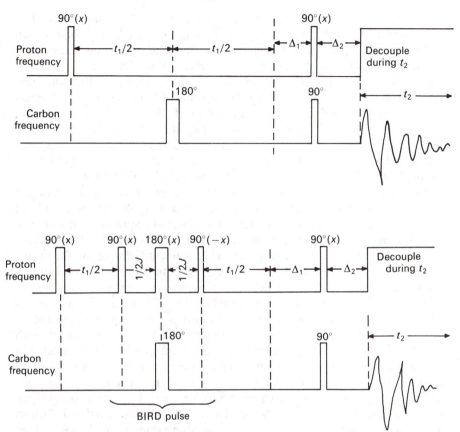

Fig. 8.9 Pulse sequences used to produce ^1H–^{13}C hetero COSY spectra. Upper sequence: the values of Δ_1, Δ_2 are made equal to $1/2J$, where J is the ^1H–^{13}C coupling constant. Lower sequence: that used to produce Fig. 8.10. Δ was made 3.6 ms, corresponding to an average one-bond C–H coupling of 140 Hz. The value of J assumed for the BIRD pulse was 7 Hz.

the protons in t_2, while accumulating the ^{13}C FID. The sequences used illustrate also how broad-band decoupling of the protons may be achieved simultaneously by using a specially adapted pulse sequence. The pulse sequences needed are illustrated in Fig. 8.9, and are shown in order of increasing complexity so that the experimental technique can be built up in stages. The first sequence shown produces ^1H–^{13}C connectivities via the one-bond coupling. The first 90° pulse at the ^1H frequency produces transverse y magnetization. This then evolves with t_1 and the position of the magnetization of a given proton depends upon its chemical shift. Those protons attached to carbon will also have two magnetization components, which precess at different frequencies, and these are refocused by inverting the carbon magnetization half-way through the t_1 period.

In principle, then, the proton magnetization at the end of t_1 is a function of t_1, and this can be transferred into the carbon magnetization by a second proton 90° pulse, which places the y magnetization into the z direction and permits polarization transfer. Unfortunately, the inversion of the ^{13}C nuclei also causes the proton lines of the ^{1}H–^{13}C doublet to be of opposed phases with zero resultant. It is therefore necessary to wait a period of Δ_1 for the two components to come into phase before the 90° pulse is applied. The delay Δ_1, which is fixed, is made to be $1/2J(CH)$. A 90° pulse is applied at the carbon frequency at the same time as the second proton pulse, which creates transverse carbon magnetization and produces an output. It is convenient to remove the multiplicity due to coupling to the protons from the carbon signals and this is done broad-band decoupling. However, again because of the different phases of the multiplet signals, it is necessary to introduce a refocusing delay Δ_2, which often is of the same length as Δ_1, before decoupling and signal acquisition are initiated. The experiment is repeated for many values of t_1 and a stack of FIDs obtained, which contain both carbon and proton chemical shift information for all the C–H pairs in the molecule. The pulse sequence is tuned to $^{1}J(CH)$ because of the value chosen for Δ and which also suppresses any responses due to the much smaller two- and three-bond couplings. It follows that, in this type of hetero COSY spectrum, quaternary carbon atoms will not give a response.

A spectrum obtained in this way for D-amygdalin is shown in Fig. 8.10, with the ^{13}C spectrum along the horizontal axis and protons on the vertical. The lack of significant response for the phenyl quaternary carbon and CN carbon atoms is evident. It is also possible to read off the carbon chemical shifts corresponding to the now known proton assignments. In order to simplify the responses, the actual pulse sequence used was more complex than that illustrated in the upper part of Fig. 8.9, and was conceived to decouple the proton resonances also. This sequence is shown in the lower part of the figure and gives an idea of the way it is possible to manipulate spin systems to obtain particular objectives. Three pulses are added to the proton sequence and one to the carbon sequence, with a particular relationship between them, which is called a BIRD (bilinear rotation decoupling) pulse. The new carbon pulse, however, has the effect of the 180° pulse in the original sequence, so that this can be omitted. This is perhaps a somewhat roundabout way to describe the changes, but it is useful to emphasize the integrality of the BIRD pulse. Its effect is best envisaged by considering what happens to the two vicinally coupled protons in a fragment H_m ^{12}C–$^{13}CH_a$ the letters a and m serving to label the two different protons. For H_m, which is not directly bonded to ^{13}C, the 180° pulse at the ^{13}C frequency has no effect. The three proton pulses then simply invert the H_m magnetization in the middle of the t_1 period, and their average influence at H_a is zero; which is to say they are decoupled from H_a. In the case of H_a, however, the magnetization precesses at two different frequencies because of the coupling to ^{13}C. The two components separate after the first pulse and are inverted by the second, *and* their frequencies are interchanged by the

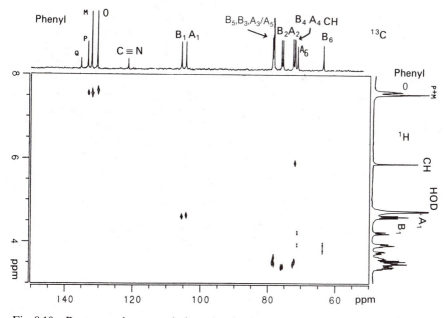

Fig. 8.10 Proton–carbon correlation plot for hydrogen and carbon atoms that are directly bonded in the molecule D-amygdalin. (From Ribiero (1990) *Magn. Reson. Chem.*, **28**, 765–73; copyright (1990) John Wiley and Sons Ltd, reprinted with permission.)

inversion of the ^{13}C magnetization caused simultaneously. By the time of the third pulse, the two components have returned to their starting position after the first pulse, and the third pulse then simply returns them to the position they had before the BIRD pulse was applied. H_a is thus not affected by the BIRD pulse and behaves exactly as in the basic sequence already described. Since the only protons that contribute to the proton spectrum in a carbon detected spectrum are those attached to ^{13}C, then no coupling to vicinal protons should be evident in the 2D spectrum obtained in this way. This will be seen to be so in Fig. 8.10, except for the protons 6 and 6′ of each ring where, because the 6, 6′ chemical shift of 150 Hz is similar to the $^1J(CH)$ value assumed, the decoupling breaks down and multiplet structure is obtained.

Confirmation of the assignments made from Fig. 8.10 can be obtained by repeating the experiment but with a value of Δ_1 and Δ_2 tuned to the much smaller two- and three-bond CH coupling constants of about 6.3 Hz. The upper sequence of Fig. 8.9 can be used for this, but there is then interference from the effects of the direct bond coupling. This is removed by the device of putting a BIRD pulse tuned now to $1/2^1J(CH)$ in the middle of the Δ_2 delay period. The resulting spectrum is shown in Fig. 8.11 and provides a great deal of extra data, which gives unequivocal assignments. For instance, the B_1H to A_6C and A_6H to B_1C responses show the connectivity between the two rings.

We have thus used a series of 2D experiments to make a full assignment of the proton and carbon spectra of D-amygdalin and obtain unequivocal evidence about its structure, despite the very congested and unpromising appearance of the 1D spectrum as a source of useful information. We have described five 2D techniques. A recent review (see bibliography) lists some 40 pulse sequences, not including the ZQCOSY sequence used here, and there are said to be perhaps 500 variations of 2D spectroscopy proposed currently. There exists, then, a whole battery of techniques that permit the successful examination of molecules much more complex than D-amygdalin, and 2D NMR is being extensively used by biochemists to understand the properties (and, indeed, structures in solution) of many molecules found to regulate the behaviour of living systems.

8.6   2D chemical exchange spectroscopy

Chemical exchange has perforce to be a homonuclear process and is studied in two dimensions by the pulse sequence shown in Fig. 8.12. If we have two

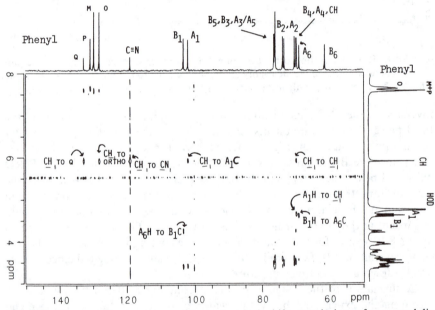

Fig. 8.11 Heteronuclear proton–carbon chemical shift correlation of D-amygdalin optimized for ^2J(CH) and ^3J(CH) long-range couplings (6.25 Hz). Direct responses due to ^1J(CH) were suppressed with the use of a BIRD pulse. Major inter-residue, long-range responses have been labelled. Additional intra-residue, long-range correlations include:
Ring B: BH2 → BC3; BH3 → BC2; BH3 → BC4; BH4 → BC5 and BH6 → BC4.
Ring A: AH2 → AC1; AH2 → AC3; AH3 → AC2; AH3 → AC4; AH4 → AC3; AH4 → AC5; AH4 → AC6; AH5 → AC3.
Aryl ring: *ortho* H → *para* C; *meta* H → quaternary; *meta* H to *ortho* C; *para* H → *ortho* C; *para* H → *meta* C. (From Ribiero (1990) *Magn. Reson. Chem.*, **28**, 765–73; copyright (1990) John Wiley and Sons Ltd, reprinted with permission.)

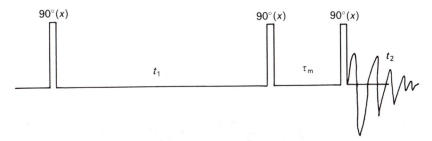

Fig. 8.12 The pulse sequence used for 2D exchange spectroscopy (EXCSY) and for NOESY. τ_m is a short mixing period where magnetization transfer occurs between non-equilibrium spin populations by either chemical exchange or through-space relaxation NOE effects.

uncoupled spins with different chemical shifts that undergo slow exchange, then the pulse sequence affects them as follows. The first 90° pulse places the magnetization in the xy plane and the two components then precess at their individual frequencies for a time t_1. At the end of t_1, they will each have a particular orientation, and in general both will have x and y components of magnetization. The second pulse then turns this magnetization into the xz plane, where the y components continue to precess about the z axis. A short mixing time (τ_m), typically 0.05 s, is then given during which time there will be an exchange of magnetization. However, the z magnetization of the two components depends upon t_1; indeed, for some ranges of t_1 one of the two components will be inverted and the exchange will cause quite marked changes in magnetization. Thus at the end of τ_m we have the two component frequencies modulated by the exchange process as a function of t_1, and this gives cross-peaks in the transformed 2D trace. A FID is produced by moving the z magnetization back into the xy plane at the end of τ_m. We should also note that, in non-exchanging molecules, if there exists through-space relaxation, then there can also be an exchange of magnetization through the NOE effect, since the second pulse produces non-equilibrium spin distributions. The same pulse sequence is thus used also for NOESY experiments as mentioned above. The acronym EXCSY is often used for the exchange experiment.

As an example of EXCSY, we show the study of ligand exchange on *cis*- and *trans*-$ZrCl_4[(MeO)_3PO]_2$ and how this is affected by pressure. The exchange is complex because it occurs between three species, the *cis* and *trans* complexes and free ligand:

$$cis\text{-}ZrCl_4L_2 + L^* \underset{}{\overset{k_c}{\rightleftharpoons}} cis\text{-}ZrCl_4LL^* + L$$

$$trans\text{-}ZrCl_4L_2 + L^* \underset{}{\overset{k_t}{\rightleftharpoons}} trans\text{-}ZrCl_4LL^* + L$$

$$cis\text{-}ZrCl_4L_2 \underset{}{\overset{k_i}{\rightleftharpoons}} trans\text{-}ZrCl_4L_2$$

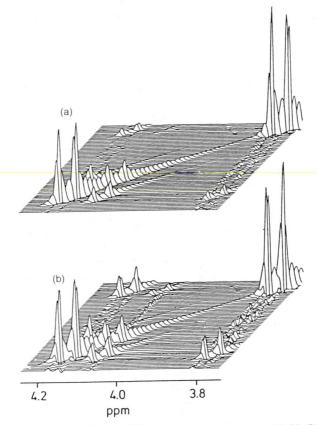

Fig. 8.13 The 400 MHz 2D ^1H EXCSY spectra of *cis*- and *trans*-ZrCl$_4$ [(MeO)$_3$PO]$_2$ in the presence of excess (MeO)$_3$PO, at two different pressures, (a) 0.1 MPa, (b) 198 MPa, demonstrating the effect of pressure on the intramolecular and intermolecular exchange pathways. (From Merbach *et al.* (1990) *Helv. Chim. Acta*, **73**, 199, with permission.)

If we are to study such a system by normal 1D techniques, it is necessary that all the species can be seen separately and line broadening followed for them all as a function of temperature or pressure as required. In this case, at temperatures near 236 K, the proton spectrum in chloroform solution consists of three methyl doublets due to the three types of (MeO)$_3$PO molecules present in the free and complexed forms. The doublet splitting is due to coupling to the ^{31}P and $^3J(^1H-^{31}P)$ is 14.0 Hz in the free ligand and 11.7 and 11.8 Hz in the *cis* and *trans* complexes respectively. As the temperature is raised, the doublets of the *cis* and *trans* complexes coalesce, while the free-ligand signal remains resolved and broadened. The isomerization reaction thus occurs at the fastest rate, and this prevents us from discovering which of the other two reactions takes place, or if both take place then what are their relative rates. This problem, impossible to solve by 1D methods, can nevertheless be solved by a 2D method. Two 2D traces obtained at different pressures are shown in Fig. 8.13, presented this time as stacked spectra rather than as contour plots. The 1D spectrum appears on the diagonal, which has a length of 0.45 ppm. The intense doublet at high field (on the right of the plots) is the free ligand, the doublet at the extreme other end of the diagonal is due to the *trans* isomer and the smaller doublet of the *cis* form is to high field of this. At ambient pressure there are strong cross-peaks connecting the two complex isomers, as would be expected. Weaker cross-peaks connect the free ligand and *trans* isomer, and the free ligand and *cis* isomer are just discernibly connected also. The cross-peak intensity is a function of the rate of exchange, so that this can be measured from these traces. Increase in pressure produces an increase in intensity of all the cross-peaks, so that the rate of exchange increases with pressure, and the mechanism of exchange with free ligand has to be associative. The increase is relatively small for the isomerization reaction, indicating a small reduction in volume during the ligand rearrangement.

8.7   3D spectroscopy

It will come as no surprise that the techniques of 2D NMR can be combined to produce experiments into which further dimensions are introduced. Possibilities include NOESY–COSY spectra or correlations between ^{15}N in ^{15}N-enriched molecules with the proton–proton COSY spectrum. While such techniques are in their infancy, they should assist in improving the resolution of the spectra of complex molecules, and are likely to take up an increasing amount of the time of NMR spectroscopists.

9 Magnetic resonance imaging and biomedical NMR

Magnetic resonance imaging has become a very useful diagnostic tool for the medical profession and is used to obtain detailed pictures of cross-sections of people's anatomy, when many abnormalities can be detected. While the initial development was concentrated in medicine, the technique has now been developed both for technological purposes and for the microscopic examination of specimens. We will describe here two of the many ways in which images can be produced and then look briefly at the various applications.

9.1 Producing an image

The spectra that we have been considering so far contain resolved structure, which arises from chemical differentiation of the nuclei observed and which are obtained in an accurately defined, homogeneous, fixed magnetic field. If, on the other hand, we were to use a magnetic field that had a gradient in one direction and a sample with one type of nucleus, say ^{1}H in H_2O, then the frequency of the water resonance would vary across the sample, and the output would thus contain information about the gradient and the disposition of the sample within it; in other words, the gradient encodes spatial information for us. Since the human body contains a large proportion of water, it became evident that, if one could map its spatial distribution, then one might be able to investigate the nature of the soft body tissue. Water is present as about 55% of body weight and the proportion varies widely in different parts of the body, as does the water mobility, and so its relaxation rates. In general, in body fluids $T_2 < T_1$, and techniques have been devised in which the image contrast can be varied by weighting either via T_1 or T_2 relaxation times.

Since an image is three-dimensional and the display of an image has to be two-dimensional, it is necessary to define a plane whose two-dimensional image is to be reproduced. This is done by subjecting the object to be imaged to a field gradient (and so frequency gradient) in the z direction only. A 90° pulse is applied, but with a fairly long duration, so that the nuclei are in resonance and swing into the xy plane over only a short distance. For instance, if the field gradient $dB_0/dz = 5\,\mu T\,mm^{-1}$, then the frequency change for ^{1}H that occurs for a 2 cm displacement along the z axis is 4150 Hz, and a pulse of the order

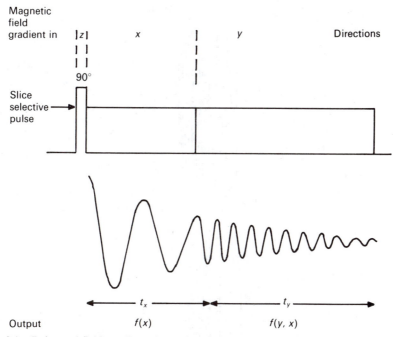

Magnetic field gradient in | z | x | y | Directions

90°

Slice selective pulse

t_x t_y

Output $f(x)$ $f(y, x)$

Fig. 9.1   Pulse and field gradient timing for obtaining a proton density image of a slice of a subject.

of 240 μs long will define a slice of the object 2 cm thick. The z field gradient is switched off at the end of the pulse and, say, gradient x is then switched on. The water in the 2 cm slice then produces a signal, the frequency of which depends upon the x coordinate only. The magnetization evolves with time and, after some time t_x, that at each x coordinate will have a particular phase, which is a function of t_x and x coordinate. The x gradient is then switched off and replaced by a y gradient. The output frequencies now depend upon the y coordinate but with phase determined by the x coordinate. The output is collected in the time domain t_y. Outputs are collected for several values of t_x and the data then subjected to a two-dimensional Fourier transform, to give a map of proton density in the slice originally defined by the combination of pulse and z gradient. The timing of the experiment is summarized in Fig. 9.1 and the way the signal phase changes in Fig. 9.2. Extra pulses can be added to this sequence to provide images whose intensity variation is a function of both relaxation time and local concentration. Thus, if the first pulse is a 180° pulse followed by a delay, then the output is T_1-weighted. On the other hand, if a Carr–Purcell sequence is used, then the output can be T_2-weighted.

The relaxation times of water protons in living material are rather short and there is a danger that much signal intensity will be lost using the method

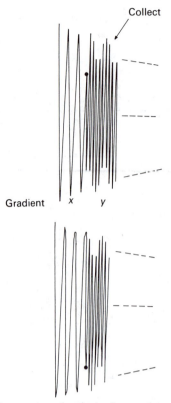

Fig. 9.2   Two outputs from an imaging device for two image elements. The low-frequency signal is produced under gradient x and, because the frequencies are different, finishes at different parts of the cycle when the x gradient is switched to y. The starting point of the signal defining the y frequency thus depends upon the x frequency also.

described. A different approach is shown in Fig. 9.3 which uses the refocusing property of the spin-echo experiment to improve the signal intensity. A non-selective 90° pulse is applied initially and places all the magnetization in the xy plane. A gradient is applied along one axis for a short time and there is dephasing of spins due to this and to relaxation effects. A gradient is applied simultaneously along a second axis, but a different magnitude of gradient is used for each pulse/gradient series and encodes the spatial information. The slice selection is done using a selective 180° pulse. This is not a rectangular pulse but has a super-Gaussian shape which, it will be remembered from Chapter 5, has a square shaped frequency transform and represents only a narrow band of frequencies. The third gradient is switched on before, during and after this pulse. Dephasing of magnetization occurs but the fraction that has the same frequency as the selective pulse is refocused while the rest is not and this defines the slice. The gradient slope is varied to ensure perfect refocusing.

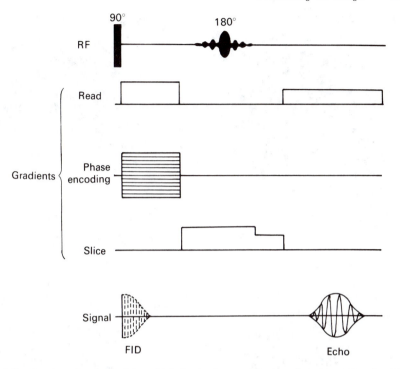

Fig. 9.3   A spin-echo method for obtaining an image. The time between pulses is of the order of 20 ms. (From Lohman *et al.* (1987) *J.Exptl. Bot.*, **38**, 1713, with permission from Oxford University Press.)

The selective 180° pulse also of course, inverts the vectors dephased by the read gradient and these are then refocused by the reapplication of the read gradient. A series of echos is obtained whose form depends upon the phase encoding gradient and as in the last example, these can be double Fourier transformed to give the spatial distribution of the water in the selected slice.

Selective pulses of the type described are now being much used in two dimensional spectroscopy and indeed pulse shaping, to obtain specific frequency responses and so particular behaviour of the nuclei in a sample, is currently being much studied.

Relaxation-time weighting has been found to be particularly useful because the contrast between different types of tissue can be markedly altered in a way that aids interpretation of the images. In particular, the relaxation properties of tumour tissue differ significantly from those of normal tissue, and tumours can be located with precision by taking a series of images with different relaxation weightings. Figure 9.4 shows an example. The production of an image takes an appreciable time if high definition is required, and this means that imaging is confined to the more static parts of the body. Such processes as the beating of the heart can, however, be observed if the data acquisition is synchronized with

BRUKER MEDIZINTECHNIK TOMIKON S 50

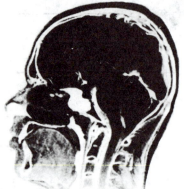

RACE
TR: 2.2s, TE: 23...46ms, S: 4mm, Ts: 9min

(a)

BRUKER MEDIZINTECHNIK TOMIKON S 50

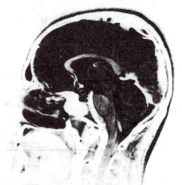

RACE
TR: 2.2s, TE: 69..184ms, S: 4mm, Ts: 9min

(b)

BRUKER MEDIZINTECHNIK TOMIKON S 50

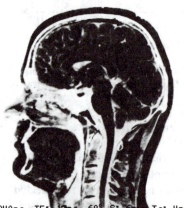

TR: 240ms, TE: 12ms, 60° S: 6mm Ts: 4min

(c)

BRUKER MEDIZINTECHNIK TOMIKON S 200
MENINGIOMA AFTER Gd-DTPA (f, 65y)

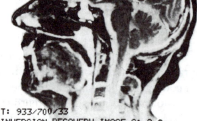

T: 933/700/33
INVERSION RECOVERY IMAGE S: 2 8mm

(d)

Fig. 9.4 Images of sections through human brains and through human body cavity showing the liver. The radiographs marked S50 were taken at a magnetic field strength of 0.5T and those marked S200 at 2.0T. The quantities given at the foot of each image are TR = time between each pulse sequence, TE = echo time interval (if several times are given then several echos have been collected and added), S = slice thickness and Ts = total time needed to collect image data. The technique used to obtain the images is related to the echo technique depicted in Fig. 9.3 but is capable of much more variation. Images (a), (b) and (c) are sections of a human head obtained using different timing sequences which give different relaxation weightings and so different contrast in the

Continued

BRUKER MEDIZINTECHNIK TOMIKON S 200
MENINGIOMA AFTER Gd-DTPA (f, 65y)

BRUKER MEDIZINTECHNIK TOMIKON S 50
LESION IN THE RIGHT LOBE OF THE LIVER
(m, 56y)

R   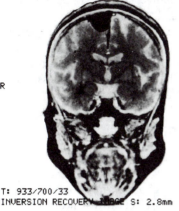   L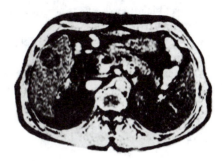

T: 933/700/33
INVERSION RECOVERY IMAGE S: 2.8mm

TURBORARE
TR: 730ms, TE: 25ms, Ts: 11s

(e)   (f)

BRUKER MEDIZINTECHNIK TOMIKON S 50
LESION IN THE RIGHT LOBE OF THE LIVER
(m, 56y)

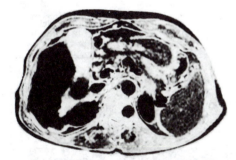

AFTER Gd-DTPA
TR: 50ms, TE: 7ms, 70°, Ts: 12s

(g)

images. Images (d) and (e) are also sections of a human head, of a patient suffering from a brain tumour. The patient had been given a paramagnetic gadolinium complex which concentrates in the tumour and reduces the relaxation time of the tumour tissue. By including an inversion recovery sequence in the imaging pulse sequence, the image can be T_1 weighted and the contrast between the tumour and normal tissue increased. Images (f) and (g) show a section through a human liver, without and with gadolinium and show the very marked changes in contrast which can be obtained. Note that in all these images the patient's right side is on the left of the image. (Examples supplied by Bruker Medizentechnik and reproduced with permission.)

the heart action. In order to produce an image, the patient has to be placed in a strong magnetic field and subjected to rapid changes in field gradient and to strong but short radiofrequency pulses. This has been found not to have any noticeable adverse effects, and an advantage of the method is that it is completely non-invasive and has none of the drawbacks of X-ray imaging.

At high enough magnetic field it is possible to resolve the resonances of chemically shifted protons, namely fatty material and water, and the possibility now exists of obtaining spectra from localized parts of the body. Nor is this type of work now limited to the proton, as ^{31}P and ^{23}Na can also be investigated.

9.2   Others uses of imaging

The measurement of the distribution of fluid throughout a sample is of interest in a number of non-medical fields and, since the objects to be measured are generally rather smaller than a human being, the size and cost of the apparatus are more reasonable. Imaging has, for instance, been used to detect the position of material on a chromatography column, which otherwise was difficult to detect. It has a number of technological uses, such as the study of water in food materials or the cooking of foods, where the non-destructive nature of the technique is useful.

The diffusion of fluids is solids can also be studied. Figure 9.5 shows images of the cross-section of a polystyrene rod that has been immersed in toluene and

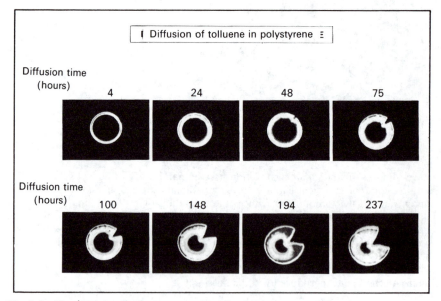

Fig. 9.5   The ^1H images of a cross-section of a polystyrene rod immersed in toluene at different times of immersion. Stress cracks form as a result of solvent-induced swelling. (From Bruker Report 1/1985, with permission.)

examined at intervals to find how deep the solvent had penetrated. The formation of a stress crack is also observed. The images shown are not relaxation-weighted and give the full spin density. If, however, the images are produced by a Carr–Purcell pulse sequence and the later echoes processed, these contain signals only from those solvent protons which have longer values of T_2, i.e. the more fluid interstitial molecules, and it is possible to separate these from those more intimately associated with the polymer matrix. The state of water in many materials is also of interest. A topical problem is the degree of damage being sustained by forests due to acid rain or other factors. Classical methods for analysis of wood are tedious and give only local information on the state of a tree. Imaging of samples can, however, show up the ring structure and differentiate live from dead wood. It has thus become possible to consider field studies of forests and compare the state of the tree crowns – a common means of observing the onset of damage – with water content and distribution within growth rings.

Imaging is also being developed for microscopic samples, since these can be examined non-destructively, and the technique has been named NMR

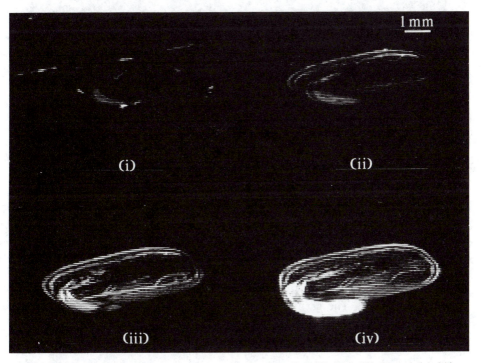

Fig. 9.6   Images of a germinating mung bean taken (i) 2, (ii) 17.5, (iii) 29 and (iv) 44 h after the start of germination, with the slice displayed centred approximately on the embryonic axis. The in-plane resolution was $70 \times 70\,\mu m$ with a slice thickness of 590 μm. The emergence of the root and the development of structure within the cotyledon are visible. (From Connelly *et al.* (1987) *J. Exptl. Bot.*, **38**, 1713, with permission of Oxford University Press.)

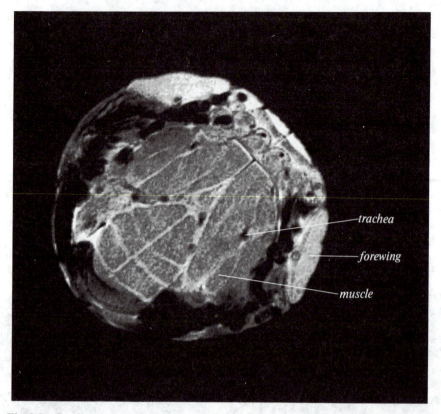

Fig. 9.7   Gradient echo image of a butterfly pupa. In-plane resolution 36 × 36 μm, slice thickness 320 μm. (From Bruker notes, *NMR Microscopy*, with permission.)

microscopy. The resolution attainable is of the order of 30 × 30 μm in plane with a plane thickness of 300 to 500 μm. This is less than available with optical microscopy, but different things are imaged by the two techniques and, more importantly, NMR is non-invasive and the sample needs little preparation. Indeed, it will probably be possible to study sections of living, whole, small plants by NMR. Water uptake can be monitored, for instance by a germinating bean seed, as demonstrated in Fig. 9.6, or by adding a relaxation agent and monitoring the loss of signal. The gradient-echo image of a section of a butterfly pupa illustrated in Fig. 9.7 is also fascinating.

9.3   *In vivo* biomedical NMR

In addition to the mobile protons in a living organism, there are molecules containing ^{13}C and ^{31}P nuclei that can also be used for study of how the chemistry of life operates. Early experiments with whole, anaesthetized, small animals in large-bore magnets were promising, but signals were obtained from all parts of the animal. It quickly became clear that it would be of most use if signals could be localized to a specific known organ. One approach is to create a specially contoured magnetic field that has the correct value for resonance over only a small, controllable volume. Alternatively, a coil may be placed on the body surface. This coil produces a radiofrequency field that penetrates below the surface by approximately its own radius and so can stimulate responses from nuclei inside the body. This is called 'topical NMR'. It is possible, for instance, to follow the ^{31}P signals of nucleotide phosphates and inorganic phosphate in muscle and monitor how these respond to different conditions imposed on the muscle or to various disease conditions. Figure 9.8 demonstrates the effect on the ^{31}P spectrum of a subject's arm of applying a tourniquet. The normal spectrum shows peaks for the three phosphorus atoms in adenosine triphosphate ATP (I, II and III), for phosphocreatine (IV) and for inorganic phosphate (V). Application of the tourniquet leads to oxygen starvation and breakdown of organic phosphate to inorganic phosphate. The effects of exercise

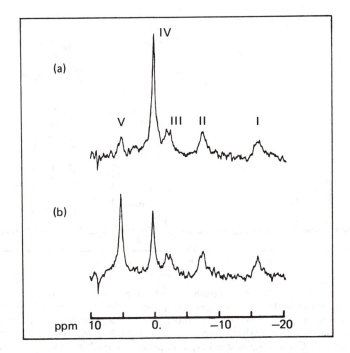

Fig. 9.8 The ^{31}P topical NMR spectra obtained at 32.5 MHz using a single-turn coil placed on the surface of a subject's arm, in a contoured magnetic field. (a) The normal spectrum (64 2-s scans). Peak assignments are given in the text. (b) The same subject 50 min after the application of a tourniquet to the upper arm. (Reproduced with permission from Gordon (1981) *Eur. Spectrosc. News*, **38**, 21.)

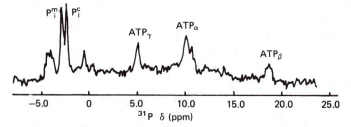

Fig. 9.9 The ^{31}P NMR spectrum of perfused rat liver cells. The two inorganic phosphate peaks P_i arise from phosphate in the mitochondria and cytosol, the latter being sensitive to the external pH while the former changes very little. The three lines marked ATP are from the three phosphorous atoms in adenosine triphosphate. The pH difference has been increased by adding valinomycin, which is an ionophore for K^+. (Reproduced with permission from Shulman *et al.* (1979) *Science*, **205**, 160; copyright (1979) by the AAAS.)

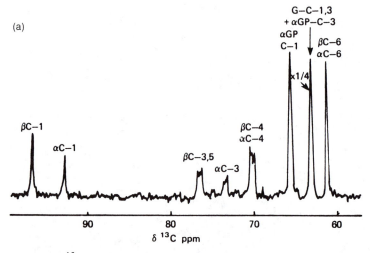

Fig. 9.10 (a) The ^{13}C NMR spectrum at 25°C of rat liver cells from a normal rat, accumulated between 18 and 35 min after the introduction of 22 mM [1,3-^{13}C] glycerol. Glucose is labelled strongly at the 1,3,4 and 6 carbon positions of the α and β anomers of glucose. Note that the carbon–carbon spin coupling has split the lines from C3 and C4 of glucose. G-C-1,3 is the labelled glycerol peak, and αGP is L-glycerol 3-phosphate. The carbon–carbon coupling splits the glucose resonances apparently into triplets because these are superpositions of the singlet due to the molecules containing only one ^{13}C between the lines of the doublet due to those containing two ^{13}C atoms. (b) Selected steps in the pathway of glucose metabolism. (Reproduced with permission from Shulman *et al.* (1979) *Science*, **205**, 160; copyright (1979) by the AAAS.)

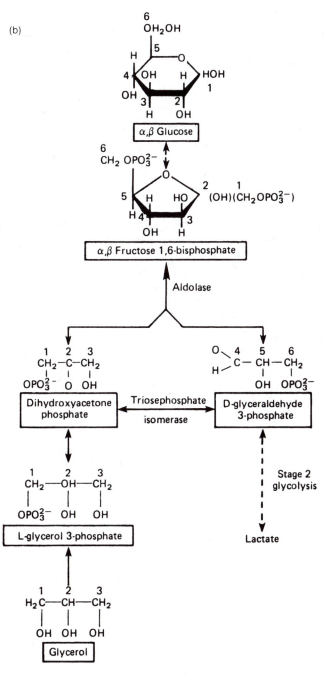

(b)

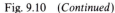

Fig. 9.10   (*Continued*)

on muscle chemistry, the spectra of skin tumours or the brain can all be obtained in this way. It is, for instance, possible to follow changes during treatment of a baby suffering from birth asphyxia using such techniques.

Other biological studies involve rather heterogeneous samples, which the NMR spectroscopist would tend to avoid. The advent of high-field spectrometers has, however, meant that worse resolution can be accepted, and results are now being reported for samples such as cell suspensions or living perfused organs. A ^{31}P spectrum of perfused rat liver is shown in Fig. 9.9 and is of interest because two sorts of inorganic phosphate are visible, inside and outside the cells. This arises because there is a pH difference between the two regions and there is a pH-dependent shift of ^{31}P due to the protonation reaction

$$H_2PO_4^- \rightleftharpoons HPO_4^{2-}$$

The chemical shift is a weighted average of the shifts of the two ions (0.4 and 3.0 ppm), which undergo rapid exchange, and whose relative populations vary with pH. Metabolic pathways can be followed in similar samples using ^{13}C spectroscopy if ^{13}C-enriched substrates are added to the sample. Figure 9.10 shows the spectrum obtained by accumulating data between the 18th and 35th minutes after adding $[1, 3-^{13}C]$glycerol to rat liver cells. The ^{13}C label appears in the 1, 3, 4 and 6 carbons of glucose and peaks due to L-glyceryl 3-phosphate are also visible. Certain of the glucose carbons show carbon–carbon spin coupling, since enrichment leads to there being more than one ^{13}C atom in each molecule.

10 High-resolution solid-state NMR

We have already mentioned in Chapter 4 that, in the solid state, the relaxation time T_1 is long due to the lack of modulation of the dipole–dipole interaction and T_2 is short due to mutual spin flips occurring between pairs of spins. In a static solid, each nucleus produces a rotating magnetic field as it precesses in the applied magnetic field, and this can cause direct exchange of energy between nuclei. The lifetimes of the spin states are thus reduced and so T_2. In addition, each spin has a static field component that influences the Larmor frequencies of its neighbours. An individual nucleus will experience the fields of several neighbours, but their spin directions will vary randomly, so that there will be a range of frequencies that will add to the line broadening due to the rapid rate of relaxation. Finally, particularly for the heavier nuclei, including ^{13}C, there may exist a chemical shift anisotropy, which will also contribute to the broadening, assuming that the sample is a powder or a glass and not a single crystal, because the chemical shift varies with orientation relative to the B_0 direction. Thus solid materials, particularly if they contain nuclei with high magnetic moments such as ^1H or ^{19}F, will have broad, structureless resonances, which will not permit the type of investigation that we have shown can be carried out in the liquid phase. This state of affairs has proved a challange to the NMR community, who have over the last decade found means to render ineffective the apparent physical restraints to the spectroscopic examination of solids at high resolution.

Before discussing this work in detail, however, it is necessary to mention two useful aspects of the broad lines. In the first place, because the broadening is determined by the dipole–dipole interactions, it is sensitive to the distance separating interacting spins. The spectrum of a solid can thus be used to obtain internuclear distances, which in the case of the proton are difficult to obtain by other means. The molecules studied must be static in the solid state and must be sufficiently simple that the resonance width can be interpreted in terms of a single, principal distance. If the molecules reorient in the solid in some way, then this modulates the internuclear interaction and its magnitude is reduced, and this is the second property that proves useful. If the linewidth of a solid material is measured as a function of temperature from very low temperatures, it is often found that there are quite rapid changes in linewidth at certain transition temperatures. These mark, if the temperature is being

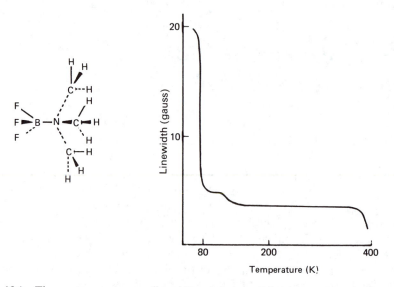

Fig. 10.1   The proton resonance linewidth of the methylamine protons in the solid adduct $(CH_3)_3N \rightarrow BF_3$ as a function of temperature. (After Dunnell (1969) *Trans. Faraday Soc.*, **65**, 1153, with permission.)

gradually increased, the onset of motion within the solid lattice. Figure 10.1 shows an example for the solid complex adduct of boron trifluoride with tri-methylamine, $Me_3N \rightarrow BF_3$. The ^1H resonance linewidth below 80 K is 85 kHz (1 gauss is equivalent to 4250 Hz in this case). Heating from 68 to 103 K reduces the linewidth to about 21 kHz, and this corresponds to the onset of rotation of the methyl groups around the CN bonds. Further narrowing to 13 kHz occurs on raising the temperature to 150 K owing to the onset of rotation of the whole NMe_3 moiety around the BN bond. Finally, just below 400 K, the line narrows to a few hundred hertz as the whole molecule starts to rotate and diffuse isotropically within the still solid crystal. The ^{19}F resonance can also be examined and is found to be broad only below 77 K and the BF_3 rotates around the BN axis at all higher temperatures. Of course, when the sample melts, the linewidth falls to a fraction of a hertz.

   Such studies using broad lines are, however, of relatively limited application, and solid-state NMR formed only a small part of the total NMR work that was undertaken. The situation has now changed dramatically with the application of the modern techniques to be described below. Both spin-1/2 and quadrupolar nuclei are studied, though the treatment of these two classes of nuclei is rather different, and we will have to discuss them separately. One technique is common to both, one which has revolutionized solid-state NMR more than any other, and we will describe this first.

10.1   Magic angle spinning

The magnetic field produced by a nucleus with magnetic moment μ at a second nucleus a distance r away will, in general, have a component in the z or B_0 direction, which influences the frequency of the second nucleus and also couples the two spins. The z component B_z is given by

$$B_z = (K\mu/r^3)(3\cos^2\theta - 1)$$

where K is a constant and θ is the angle between the direction of B_0 and the line joining the two nuclei. At one particular angle, shown in Fig. 10.2, B_z is zero. This is the angle for which $3\cos^2\theta - 1$ is zero or $\theta = 54°44'$. It is the angle at which all dipolar interactions disappear. In a real sample, of course, which typically will be a powder, the internuclear vectors take all possible angles θ and the trick is to make them behave as if all had this angle. This is done by mounting the sample in the rotor of a small air-driven turbine whose axis is inclined at an angle of $54°44'$ to the magnetic field direction. Means are provided to adjust the angle so as to obtain the optimum results. The turbine is rotated at high speed, and this gives all the internuclear vectors an average orientation at the rotor angle, which produces dramatic changes in linewidth. This is illustrated in Fig. 10.3. The angle has to be adjusted quite finely to obtain the best results and is accordingly called the 'magic angle', and the technique is

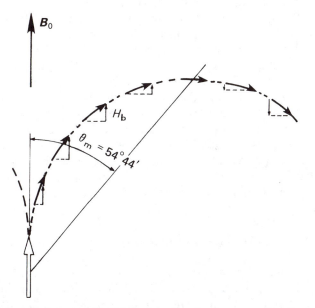

Fig. 10.2   A line of the magnetic field originating from a magnetic dipole has zero z component at a point situated on a line originating at the centre of the dipole and at an angle of $54°44'$ to the direction of the dipole. (From Bruker CXP application notes, with permission.)

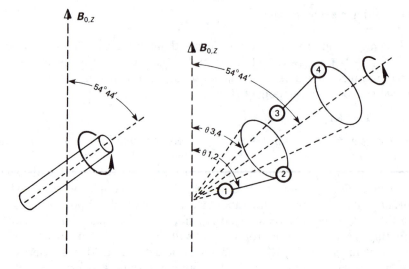

Fig. 10.3   Showing how a solid sample is mounted for magic angle spinning and how this gives the internuclear vectors an average orientation at the spinning angle. (From Bruker CXP application notes, with permission.)

'magic angle spinning', usually abbreviated to MAS. The chemical shift aniso-tropy also follows a $3\cos^2\theta - 1$ law, and broadening due to this cause is also removed by MAS. Quadrupole broadening for nuclei with $I > 1/2$ is also reduced by MAS, but the situation is more complex for the quadrupolar nuclei, as we shall see. The spinning speed is important, since, if the static linewidth of the resonance to be studied is F Hz, then the spinning speed must be greater than this if all the broadening interactions are to be nullified. Since the basic physical strengths of materials are limited, there is a limit to the centrifugal forces that they will withstand, and so a limit to the speed of spinning. Currently, the normal limiting speed is around 7 kHz, though in some laboratories speeds as high as 23 kHz can be attained. It is also useful to know the speed at which a rotor is spinning, and these are provided with marks that can be detected by a suitable frequency counter.

10.2   Spin-1/2 nuclei with low magnetogyric ratios

Common examples of compounds within this class are ^{13}C in organic com-pounds or ^{31}P in inorganic or organic phosphates. If protons are present, they, of course, cause strong dipolar broadening and their influence has to be removed. This can be achieved by high-power double irradiation, high power being required because the proton resonance will be of very large width. The ^{13}C spectra of adamantane (Fig. 10.4) show the improvements in resolution that

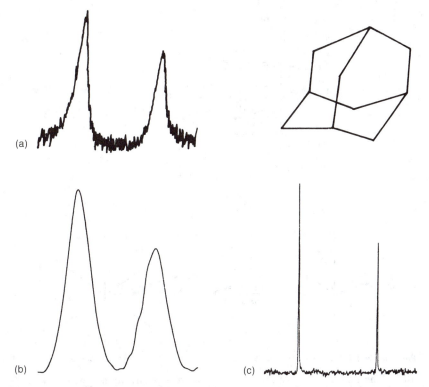

Fig. 10.4   The ^{13}C spectra of solid admantane (formula inset): (a) with MAS; (b) with high-power proton double irradiation; and (c) with both MAS and double irradiation. (From Bruker CXP application notes, with permission.)

can be obtained. The normal solid-state spectrum has a linewidth of some 5000 Hz, and the two types of carbon present in the molecule do not have resolved resonances. MAS alone reduces the linewidth to 200 Hz and enables the two types of ^{13}C to be distinguished. High-power decoupling of the protons alone also reduces the linewidth to 450 Hz in the case illustrated. Combination of the two techniques, however, reduces the linewidth to 2 Hz, which is little worse than in liquid-state samples. Obtaining solid-state spectra in this way has certain drawbacks. The rather long ^{13}C T_1 values mean that pulse rates have to be sufficiently slow so as not to saturate the resonances, and the natural insensitivity of the nucleus means that long accumulation times are needed. These disadvantages can be circumvented by a modification of the double-resonance technique, which permits the exchange of polarization between the ^{1}H and ^{13}C spins, and which is called cross-polarization (CP); when combined with MAS, the whole is abbreviated to CP-MAS.

   In order that energy exchange shall be possible between the two nuclear species, we must introduce components of motion with the same frequency for

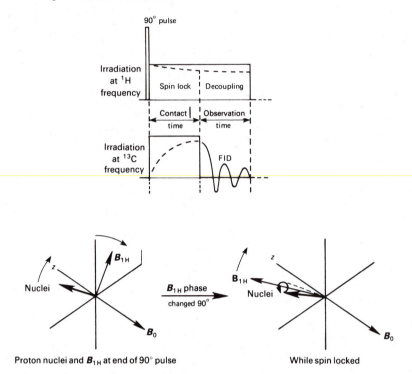

Fig. 10.5   The timing of the various events in a cross-polarization experiment. A 90° pulse at the proton frequency is followed by a long spin locking pulse changed in phase from the initial pulse by 90°, which prevents the normal proton relaxation processes. Energy is caused to flow the assembly of ^{13}C nuclei to the cooled proton nuclei by applying a long pulse at the ^{13}C frequency, which introduces a precession frequency equal to that already established for the protons. When equilibrium is reached, this field is switched off and is followed by a ^{13}C FID of enhanced intensity. (Bruker CXP Leaflet).

each. This is done as follows. Referring to Fig. 10.5, we first prepare the protons for the cross-polarization by applying a short 90° pulse, which swings the protons into the xy plane. We will call this field B_{1H}. We then simultaneously reduce B_{1H} and change its phase by 90°, so that the B_{1H} vector becomes parallel with the spins in the xy plane. The spins remain in the xy plane and stay there locked to B_{1H}, which is known as a spin locking pulse. The magnetization of the spin vector cone precesses in the xy plane at the Larmor frequency, and the distribution in the cone can be thought of as also precessing around B_{1H} at a frequency of $\gamma_H B_{1H}/2\pi$ Hz, behaving as if they were very strongly polarized in the weak B_{1H}. Reference to equation (1.3) will show that such a polarization, correct under normal circumstances for B_0, can only be attained if the temperature is very low for a field B_{1H}, and we can consider that the spin locking has cooled the spins, which can now act as an energy sink. Energy transfer is obtained by applying a long pulse at the ^{13}C frequency, B_{1C}, which has an amplitude such that the ^{13}C nuclei precess around it at a frequency of $\gamma B_{1C}/2\pi$, which is equal to $\gamma_H B_{1H}/2\pi$. The identical frequency components allow energy transfer, which follows an exponential curve, and when this has reached a maximum the B_{1C} is cut off and is followed by a ^{13}C FID of enhanced intensity. The B_{1H} field remains on during this time and provides the decoupling of the protons from the carbon nuclei. The experiment can be repeated when the proton spins have reached equilibrium and so the effective relaxation time is the shorter one of the proton system. The cross-polarization increases the ^{13}C population difference by the factor γ_H/γ_C and so produces a useful improvement in signal strength. An example of a ^{13}C CP-MAS spectrum is shown in Fig. 10.6 for a complex molecule, deoxycholic acid, together with a partial assignment of the resonances. Deoxycholic acid forms inclusion compounds, and, when the guest molecule is ferrocene, $Fe(C_5H_5)_2$, the methyl singlets due to C18, C19 and C21 become doublets due to differentiation of the deoxycholic acid molecules in the solid lattice. The cross-polarization technique is much used with low-γ nuclei where the compound studied contains hydrogen; ^{13}C, ^{15}N, ^{29}Si, ^{31}P and ^{113}Cd are some examples of spin-1/2 nuclei studied. Quadrupolar nuclei can also benefit from the CP technique.

If the spinning speed is apppreciably lower than the width of the static resonance of the compound studied, then sidebands are produced separated by the spinning speed. Figure 10.7 shows the solid-state ^{31}P spectra of amino-methanephosphonic acid, $H_3N^+CH_2PO_3H^-$, which has a zwitterionic structure. All the three spectra shown benefited from CP, but the uppermost one (a) was from a static sample, and shows the chemical shift anisotropy of the phosphorus nucleus, which does not have axial symmetry and so has three components σ_{11}, σ_{22} and σ_{33}. MAS at 813 Hz (b) produces a group of narrow lines separated by the spinning frequency. The number of lines is reduced on increasing the spinning speed to 2950 Hz (c), and it is evident that only one does not move, so that this (arrowed) is the true resonance with the isotropic chemical shift and the remainder are spinning sidebands. It is particularly important to note that the envelope enclosing the sidebands approximates the static lineshape and so retains the form of the chemical shift anisotropy, particularly at low spinning speeds. The ^{31}P spectrum of *N,N*-dimethylamino-methanediphosphonic acid, $Me_2NH^+(PO_3H_2)_2$, is also shown for a spinning speed of 1740 Hz (d). In this case there are two resonances that are not sidebands, since the two phosphorus atoms in a single molecule are not related by symmetry due to the crystal structure. In solution, of course, only a singlet ^{31}P resonance is produced by this compound.

Such high-resolution spectra allow access to parameters such as chemical shift anisotropy and permit comparison of molecular structure in solution and crystal. In addition, they permit insoluble substances to be studied at reasonably high resolution, and investigations are being carried out into the structures of materials such as plastics or coals. The latter have been notoriously difficult to

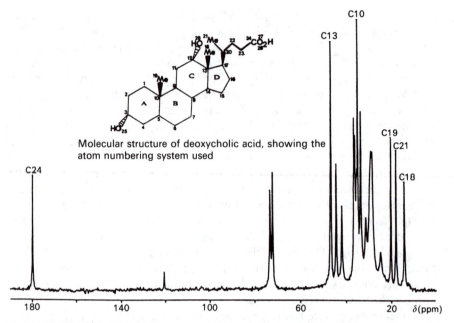

Molecular structure of deoxycholic acid, showing the atom numbering system used

Fig. 10.6 The ^{13}C CP-MAS spectrum of deoxycholic acid. The ^{13}C frequency was 50.32 MHz and that for ^{1}H was 200 MHz. Some 350 mg of sample was packed into the MAS rotor and some 800 transients were acquired. The contact time was 1 ms and recycle delay 3.5 s. (From Heys and Dobson (1990) *Magn. Reson. Chem.*, **28**, 537–46; copyright (1990) John Wiley and Sons Ltd, reprinted with permission.)

study without degrading the coal structure, and this is now possible using NMR. An example is shown in Fig. 10.8 for both a plastic and a coal, and, while the resolution for the coal is not exceptional, it has to be remembered that it is a most complex mixture of structures and that it is most useful to be able to distinguish aromatic and aliphatic resonances and perhaps to be able to do this quantitatively.

10.3   Spin-1/2 nuclei with high magnetogyric ratios

Here we are thinking specifically of the nuclei ^{1}H or ^{19}F, where the homonuclear interactions are very strong and so difficult to remove by MAS. ^{1}H spectra are

Fig. 10.7 The ^{31}P spectra of aminomethanephosphonic acid: (a) static sample; (b), (c) with MAS at 813 Hz and 2950 Hz; (d) MAS spectrum of N, N-dimethylaminomethane-diphophonic acid showing the existence of two crystallographically differentiated sites. The arrows show the centre band lines with the isotropic chemical shifts. The other lines are spinning sidebands. (From Harris et al. (1989) *Magn. Reson. Chem.*, **27**, 470, with permission.) (Copyright (1989) John Wiley and Sons Ltd, reprinted with permission.)

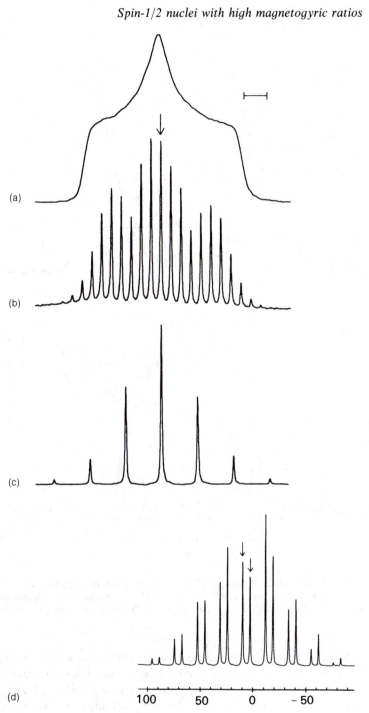

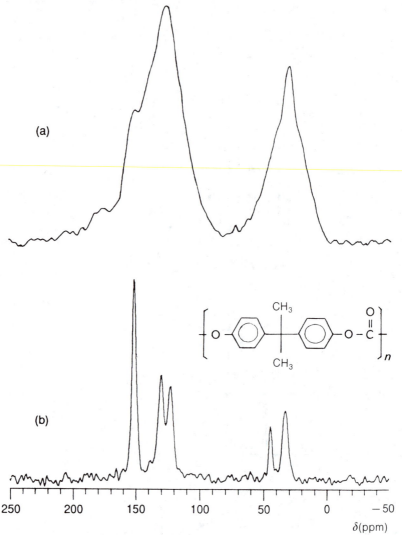

Fig. 10.8   The ^{13}C CP-MAS spectra at 15 MHz of (a) a sub-bituminous coal and (b) polycarbonate solids. (From Wind (1991) *Modern NMR Techniques and Their Application in chemistry*, Eds Popov and Hallenga, Dekker, New York, p. 186, with permission.)

the most difficult to deal with, since not only are the static linewidths very large, but the chemical shifts are small, so making big demands on the resolution ability of the system. MAS at the very highest spinning speeds and using the highest magnetic fields to maximize the chemical shift dispersion is capable of reducing a static linewidth of the order of 10 kHz to around 1500 Hz, and this can give resolvable resonances. Alternatively, the spins can be swung around by a succession of pulses so that they appear to adopt the magic angle in the rotating frame. A typical sequence, called MREV-8, is shown in Fig. 10.9 and has the effect that the magnetization is shifted quickly between the three orthogonal axes, which, as it will be remembered from section 4.3 and Fig. 4.7, are placed at the magic angle from their three-fold symmetry axis. The spins thus hop around the magic angle axis and their dipole–dipole interaction is much reduced, though the chemical shift anisotropy and heteronuclear interactions are not affected. As might be expected, then, since these will be reduced by MAS, a combination of the two methods proves very successful. This technique is

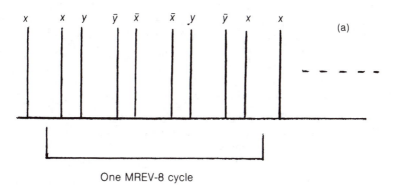

One MREV-8 cycle

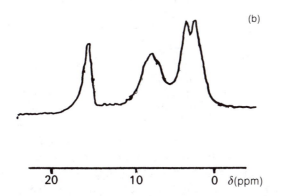

Fig. 10.9   (a) The eight-pulse MREV-8 cycle. Each pulse is a 90° pulse, which rotates the magnetization around the x or y axes in one direction or its opposite. The large spacings are double the length of the small spacings. The nuclear signal is sampled between pulses. (b) The ^{1}H CRAMPS spectrum at 200 MHz of aspartic acid, $HOOCCHNH_2CH_2COOH$. (From Bruker Report 1/1988, with permission.)

called combined rotation and multiple-pulse spectroscopy (CRAMPS). Resolution of the order of 180 Hz is possible in ^1H spectroscopy. The CRAMPS spectrum of aspartic acid, $HOOC.CHNH_2.CH_2.COOH$, is also shown in Fig. 10.9, where resonances due to the four different types of proton can easily be distinguished.

10.4   MAS of quadrupolar nuclei

Quadrupolar nuclei in the solid state usually have weak dipolar interactions with their surroundings, and this will not concern us here. Where they do exist, then a cross-polarization experiment may be possible. Quadrupole nuclei are, of course, polarized by the magnetic field but are also subject to any electric field gradient present at their position in the molecular and crystallographic environment, and which can arise from the bonding electrons and, in contrast with the liquid, from more distant electronic distribution. The nature of the interaction depends upon whether the nucleus examined has an integral or half-integral spin. As there are only effectively two important nuclei with integral spin, namely ^2H and ^{14}N, we will discuss initially only the half-integral spin nuclei. If a single crystal of a solid that contains such a nucleus is placed in a strong magnetic field at some particular orientation of a crystal axis to that field, then we have also determined the way the electric field gradient interacts with the several possible spin states. If the electric field gradient (EFG) is zero, then the energy gap between spin states is the same for any pair: that is, the energy of the transitions possible for a spin-3/2 nucleus, $3/2 \leftrightarrow 1/2$, $1/2 \leftrightarrow -1/2$, $-1/2 \leftrightarrow 3/2$, are all equal and a single resonance results. If there is an EFG, then the energy of each spin state is altered. The $1/2$ and $-1/2$ transitions move in parallel, so that the transition energy is unaltered, but the $3/2$ and $-3/2$ transitions change in the opposite direction and so reduce the energy of one transition and increase the energy of the other. The degeneracy of the three possible transitions is now lifted and we detect three resonances. This is shown schematically in Fig. 10.10. If we change the orientation of the crystal in the magnetic field, we change the orientation of the EFG tensor and so the interaction of the nucleus with the EFG. This alters the energy levels: though the $1/2 \leftrightarrow -1/2$ transition frequency is unchanged, the other two will both be affected, one being increased and one decreased. At certain orientations they will coincide with the $1/2 \leftrightarrow -1/2$ transition and, as the crystal is rotated, they will move away from this $1/2 \leftrightarrow -1/2$ centre band, reach a maximum displacement, return to the centre band, cross over, reach maximum again and, at 180° rotation, for crystals with axial symmetry again coincide with the centre band. This is known as the first-order quadrupole effect. An example of how the outer bands move as a function of rotation angle is given in Fig. 10.11, which shows the results for a single crystal of sodium nitrate placed so that its three fold axis of symmetry could be rotated to describe a plane parallel with

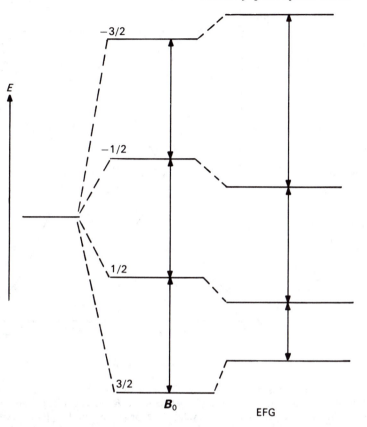

Fig. 10.10 Energy level diagram for a nucleus $I = 3/2$. The transition energies are equal if only the magnetic field B_0 is taken into account, but the quadrupole interaction causes one to become larger, one to become smaller and one, the centre band, to remain unchanged.

B_0. The maximum displacement of the satellites is proportional to the magnitude of the quadrupole coupling constant and to $3\cos^2\theta - 1$, where θ is the angle between the EFG tensor axis (V_{zz}) and the direction of the magnetic field B_0. In a powder sample, where the crystallites are arranged at random, the satellite lines have all possible positions and so smear out into the baseline, leaving only the centre band unchanged and detectable.

Because of the tensor nature of the EFG, this description is oversimplified and we have to take into account a second-order perturbation of the energy levels and so of the frequencies of all the lines. If we define a quantity v_Q as a measure of the magnitude of the quadrupole interaction, where

$$v_Q = 3e^2 qQ / 2Ih(2I - 1)$$

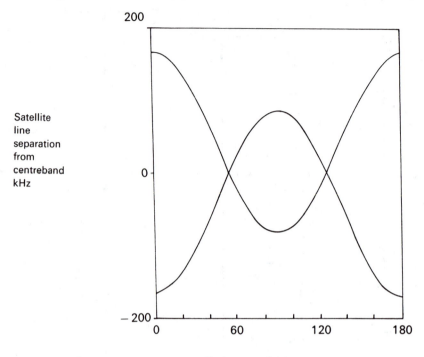

200

Satellite
line
separation
from
centreband
kHz

0

− 200

0   60   120   180

Degrees rotation

Fig. 10.11   The positions of the two satellite lines in the [23]Na spectrum of a single crystal of $NaNO_3$ as a function of angle between the B_0 magnetic field direction and the three-fold symmetry axis of the crystal. The origin of the frequency axis is the frequency of the central line. (Derived from Andrew, *Nuclear Magnetic Resonance*, Cambridge University Press (1958), with permission.)

and provided the quantity v_Q^2/v_0 is appreciable in magnitude, where v_0 is the Larmor frequency of the nuclear species examined, then the frequency of the central $1/2 \leftrightarrow -1/2$ transition also depends upon the angle θ but in a more complex way. Thus this second-order shift v_2 is

$$v_2 = (-v_Q^2/16v_0)[I(I+1) - 3/4](1 - \cos^2\theta)(9\cos^2\theta - 1)$$

for an axially symmetric field gradient with $\eta = 0$. These terms have already been defined in Chapter 4. In a powder sample, this second-order angular dependence of the line frequency imparts width, which, it has to be emphasized, is a shift effect, not a relaxation effect as in the liquid phase. The line has breadth $v_{1/2}$ and shift from the isotopic resonance frequency δ_2 of approximately

$$v_{1/2} = 25v_Q^2/18v_0 \quad \text{and} \quad -\delta_2 = v_Q^2/20v_0$$

The line shape of a powder sample is thus seen to be a valuable source of information about the nuclear environment, though the centre of the line is not at the chemical shift position and the true chemical shift has to be calculated if a precise value is required. The form of the spectrum is also a function of the spectrometer magnetic field strength, and the second-order effects become less evident as this is increased. Thus high magnetic fields are essential for the study of quadrupolar nuclei in the solid state.

The question we must next ask is, of course, will magic angle spinning reduce this linewidth? Evidently, if there is any contribution from chemical shift anisotropy or dipolar coupling, then this will be reduced or eliminated. Otherwise MAS is not so magic in the case of the quadrupolar nuclei. The second-order term, which does not vary as $3\cos^2\theta - 1$, is not eliminated by MAS, though it is reduced by a factor of about four times. Thus if $\eta = 0$, the width and shift of the resonance under MAS are approximately

$$v_{1/2} = v_Q^2/3v_0 \quad \text{and} \quad -\delta_2 = v_Q^2/4v_0$$

One has to use the term 'approximately' in describing these quantities because the line shape is complex and quite unlike the Lorentzian lines obtained with liquid samples. A theoretical line shape is sketched in Fig. 10.12 for a nucleus of spin 5/2 in a powder sample undergoing MAS with $\eta = 0$. Perhaps the most important point to note from this sketch is the fact that the true isotropic chemical shift of the resonance is not at the centre of the resonance but is to low field or high frequency always. In practice, the line shape is rounded off from the rather angular shape shown, and an estimate of the true shift can be made as being the point where the low-field side of the resonance has lost almost all its intensity. Clearly, this is approximate, and any accurate estimate requires

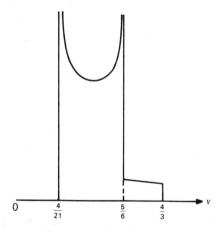

Fig. 10.12   A sketch of the line shape of a spin-5/2 nucleus in a powder sample undergoing MAS. The frequency scale is calibrated in units of $v_Q^2/2v_0$ and the origin of the shift scale is the Larmor frequency in the absence of any second-order quadrupole interaction. (From Akitt (1981) *Prog. NMR Spectrosc.*, **21**, 1; copyright (1981) Pergamon Press PLC.)

the line shape to be calculated for a particular spectrum. It is, however, a better approximation than that often encountered currently of giving the shift as the centre of the resonance. This adequately describes the position of the peak on the spectrum but is *not* its chemical shift.

The line shape of a quadrupolar nucleus in a spinning sample is strongly dependent on the angle of the turbine set relative to B_0. In fact, the best angle for reducing the second-order linewidth is not the magic angle, and it is better to choose some other angle. The actual value to use is a compromise between reducing the second-order effects and those effects, if present, that depend upon $3\cos^2\theta - 1$ terms. It is thus necessary to be able to vary the spinning angle, and this has been given the name VASS–variable angle sample spinning.

Some of the aspects of solid-state NMR that we have just discussed are summarized in Fig. 10.13, which shows the ^{27}Al spectra of the tridecameric

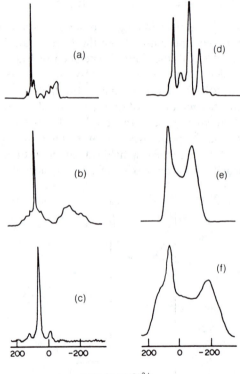

ppm from Al $(H_2O)_6^{3+}$

Fig. 10.13   The ^{27}Al solid-state spectra of the tridecameric aluminium cation at different magnetic fields, with and without sample spinning. Spectrometer frequencies were (c), (f) 39 MHz, (b), (e) 93.7 MHz and (a), (d) 129.7 MHz. Samples (e) and (f) were static, samples (a), (b) and (c) were spinning at the magic angle, and sample (d) was spinning at an angle of 75° to the magnetic field direction. (After Oldfield *et al.* (1984), *J. Magn. Reson.*, **60**, 467, with permission.)

aluminium cation already discussed in Chapter 4. This has the formula $AlO_4Al_{12}(OH)_{24}(H_2O)_{12}$, with one Al in a site of high tetrahedral symmetry and so low quadrupole coupling constant and the remaining 12 in sites of distorted octahedral symmetry and so with high quadrupole coupling constants. The solution-state spectrum contains one narrow line (AlO_4) and a very broad, usually undetectable line 58 ppm to high field (AlO_6). The solid-state spectrum obtained at low field is similar, though it must be appreciated that the AlO_6 resonance is broadened by quite different mechanisms is the two phases. As the field and v_0 are increased, the second-order broadening is reduced and the octahedral resonance becomes visible ((c), (b) and (a)). If the spinning angle is changed to 75°, then the AlO_6 linewidth is further reduced (d) and two spinning sidebands are also visible. It is remarkable to note that, in this spectrum, we have better resolution of the octahedral aluminium resonance than we do in solution. The spectra of static samples are also shown and make very evident the improvement in resolution obtained with spinning. The static spectra can nevertheless be used to calculate the quadrupole coupling constants, which are to be used to interpret the spectra of the spinning samples.

A further possibility for improving the resolution of resonances exists in the solid state which is absent for liquid samples. Since the second-order quadrupole effect produces frequency shifts, then sites with similar chemical shifts but different quadrupole coupling constants may well have resonances with centroids that are well separated. A striking example of this is shown in Fig. 10.14, which shows the ^{27}Al MAS spectrum of $CaO.3Al_2O_3.3H_2O$ obtained at 130 MHz. The crystal contains two crystallographic types of four-coordinate aluminium but which have a chemical shift of only 1 ppm, or 130 Hz in the spectrometer used. The quadrupole coupling constants at the two sites are, however, very different, so the line shapes are different and their centroids are separated by some 10 ppm, sufficient to observe the resonances separately and to compute their line shapes. Note also how the isotropic chemical shifts, shown by two marks on the chemical shifts axis near 80 ppm, coincide with essentially zero signal intensity.

10.5 Some applications

Solid-state NMR can usefully be applied to the determination of the state of the cations in the alkalide salts. These substances typically have formulae $MM'L_n$, where M and M' are an alkali metal or metals and L is a strong complexing ligand, either a cryptand or a crown ether capable of enclosing or partially enclosing the alkali-metal cation. The second alkali metal is present as the anion, M^-. The ^{23}Na MAS spectrum of the alkalide $Na^+C222.Na^-$ is shown in Fig. 10.15. C222 is the cryptand $N(C_2H_4OC_2H_4OC_2H_4)_3N$. Two resonances are observed, one with an isotropic shifts near that of the standard at 0.0 ppm, $Na^+(aq)$, and one with a shift significantly to high field, as would

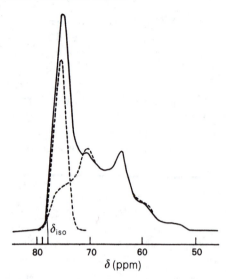

Fig. 10.14   The ^{27}Al MAS spectra at 130 MHz of CaO.3Al$_2$O$_3$.3H$_2$O showing the two superimposed signals and their computed line shapes. Note particularly the two close values of the true isotropic chemical shifts, which are marked on the shift axis and correspond with hardly any signal intensity. Note also how two types of chemically very similar aluminium are well differentiated by different quadrupole couple constants. (Reprinted with permission from Müller *et al.* (1986) *J. Chem. Soc., Dalton Trans.* 1277.)

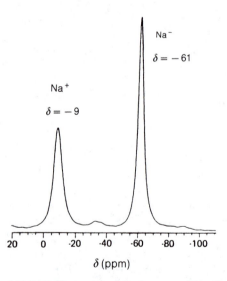

Fig. 10.15   The ^{23}Na MAS NMR spectra of the homonuclear alkalide Na$^+$C222.Na$^-$ (From Dye *et al.* (1991) *Modern NMR Techniques and Their Application in Chemistry,* Eds Popov and Hallenga, Dekker, New York, p. 291, with permission.)

be expected for the greater electron density on Na⁻. If the alkalide contains two different alkali-metal atoms, then the solid-state chemical shifts of each will indicate which is the anion and which the cation.

The examples shown up to present have all given data comparable with crystallographic data, and the work could in many cases have been achieved equally well by this older technique, though NMR is perhaps less time-consuming and can be used to follow the effect of small changes in sample conditions with relative ease. There is, however, a whole class of compounds where crystallography is of little help, such as disordered solids, glasses and amorphous substances and solids with minor but important components that are not picked up by diffraction experiments. NMR is ideal for the study of such materials and we give several examples below.

10.5.1   The setting of cement

Monocalcium aluminate, $CaO.Al_2O_3$ is the main constituent of high-alumina cement, and the ability to investigate its hydration, i.e. how it sets, is of obvious general interest and technological importance. The dry starting material contains four-coordinate aluminium, which becomes six-coordinated upon hydration. There is a chemical shift of some 70 ppm between the two types of aluminium, and so the hydration can be followed through all its stages. A typical ^{27}Al MAS spectrum is shown in Fig. 10.16, together with plots showing how the proportion of octahedral aluminium changes with time and also the amount of heat of reaction evolved with time. Setting is more rapid at high temperatures and the conversion to the octahedral form is more complete. The rate of reaction also varies quite widely at the two higher temperatures and a series of steps are observed. The initial reaction is believed to produce a mixture of phases, which covers the unreacted material and causes the reaction to slow down. This covering then suffers transformation to new phases, which expand and loosen the coating and so permit reaction to proceed again.

10.5.2   Zeolites

These substances are formed of networks of aluminosilicates that contain pores of certain fixed sizes and are active as catalysts. They are produced by crystallization from a gel formed upon mixing, say, an aluminate salt with a soluble silicate, and the structure of the zeolite formed depends upon the nature of the components used to form the gel. The ratio of silicon to aluminium present in the zeolite can be varied within certain limits and the silicon is always in excess. The solids thus contain Si–O–Si and Si–O–Al linkages but not Al–O–Al linkages, which appear to be forbidden. These substances contain two magnetically active nuclei, ^{29}Si with $I = 1/2$ and the quadrupolar ^{27}Al with $I = 5/2$, and both are used extensively in their study. The ^{29}Si spectra may contain up to five resonances, which correspond to tetrahedral SiO_4 units with

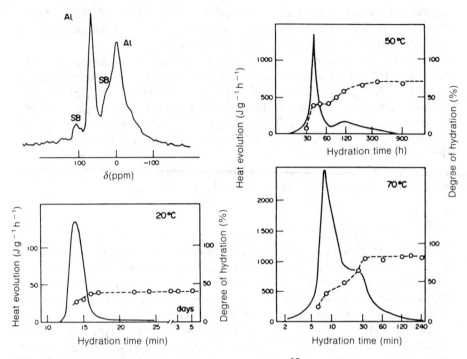

Fig. 10.16 Hydration of calcium aluminate. A typical ^{27}Al MAS spectrum is shown at the top left of the figure. The three other plots show the progress of hydration with time at three different temperatures. The full curves show the heat evolution and the broken curves show the percentage of six-coordinate aluminium formed. SB on the spectrum signifies spinning sidebands. (Reprinted with permission from Rettel *et al.* (1985), *Br. Ceram. Trans. J.* **84**, 25.)

zero, one, two, three or four attached aluminium atoms. A spectrum of zeolite Na-Y with Si/Al = 2.61 is shown in Fig. 10.17, where the resolution of the different environments is seen to be excellent. With care, the intensities are quantitative and the pattern allows the Si/Al ratio to be calculated. In natural zeolites, this ratio is always less than about 5, but materials with much lower aluminium contents can be synthesized. The series of highly siliceous zeolites called ZSM-5 with Si/Al typically 31 are a well known range of catalysts with extra stability conferred by the high silicon content. The ^{29}Si spectra of such substances are simple with essentially a single line due to Si(OSi)$_4$ units. A second material isostructural with ZSM-5 but with only a trace of aluminium and called silicalite is also known. If the aluminium content is particularly low, then the single-line ^{29}Si spectrum is found to be resolved into a group of lines, which arise from the various crystallographic sites in the as-yet uncertain structure of this material. This remarkably well resolved spectrum is also shown in Fig. 10.17. Because the aluminium content of silicalite is so low, it was argued,

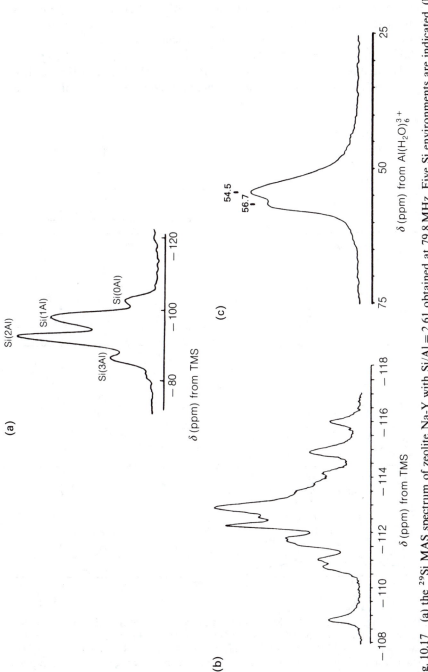

Fig. 10.17 (a) the ^{29}Si MAS spectrum of zeolite Na-Y with Si/Al = 2.61 obtained at 79.8 MHz. Five Si environments are indicated. (b) The ^{29}Si MAS spectrum of silicalite with Si/Al > 1000 obtained at 99.32 MHz. The resonances are all from SiO_4 with no directly linked Al. (c) The ^{27}Al MAS spectrum of the same sample taken at 104.2 MHz and the result of accumulating 176214 FIDs. (From Klinowski and Thomas (1985) *Adv. Catal.*, **33**, 199, and Fyfe *et al.*, (1982) *J. Phys. Chem.*, **86**, 1247, and Klinowski *et al.*, *Prog. NMR Spectrosc.*, **16**, 237; copyright (1984) Pergamon Press PLC, reprinted with permission.)

in order to be able to patent its use, that it was not a zeolite and that any aluminium was present as alumina impurity. ^{27}Al is a good, receptive nucleus and can be detected at quite low levels in solids, which are concentrated states of matter. The ^{27}Al MAS spectrum of the same sample of silicalite is shown in Fig. 10.17, and it is evident that the ^{27}Al is detectable, even though an accumulation time of over two days required to collect the 176 214 FIDs needed. The chemical shift is diagnostic for tetrahedrally coordinated aluminium, so

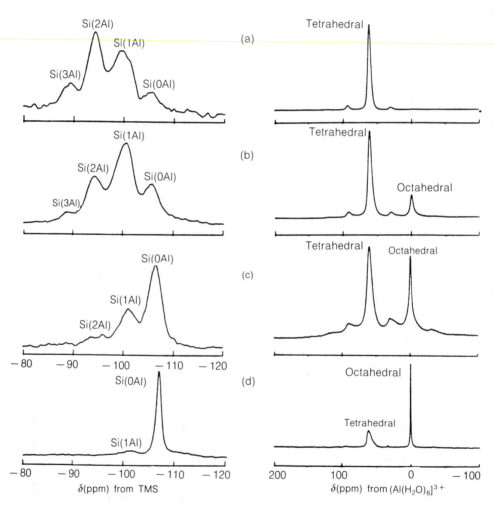

Fig. 10.18   High-resolution MAS spectra monitoring the ultrastabilization of zeolite-Y as described in the text. The left-hand spectra are ^{29}Si at 79.8 MHz and those on the right are ^{27}Al at 104.2 MHz. (From Thomas and Klinowski (1985) *Adv. Catal.*, **33**, 199, and from *Nature*, **296**, 533–6; copyright © 1982 Macmillan Magazines Ltd, reprinted with permission.)

that the aluminium is to be found within the silicalite framework, and, further, there are at least two different aluminium environments. Note, again, that it is the peak centroids that are indicated on the figure. The same type of structure has since been observed in ZSM-5 that has been thoroughly de-aluminated to reach Si/Al = 800.

An alternative approach to these catalysts is to take, for instance, zeolite Y and subject it to what is known as decationation and ultrastabilization. The ammonium form of the zeolite is subjected to heat treatment under vacuum conditions, when it loses ammonia and water. The resulting crystalline material has much greater stability than the starting material and is a good catalyst used for hydrocracking in the petroleum industry. It has a much reduced ion-exchange capacity and this indicates that aluminium has been lost from the framework, the vacancies created being reoccupied by Si. The aluminium remains but can subsequently be leached out of the solid catalyst. This ultrastabilization process has been studied by both ^{29}Si and ^{27}Al spectroscopy, as shown in Fig. 10.18. The starting material (a) had Si/Al = 2.61, four lines in the ^{29}Si spectrum and all tetrahedral Al. After calcining in air at 400°C for one hour (b), there are evidently fewer aluminium atoms linked to the silicon and some octahedral aluminium has appeared. Si/Al was calculated to be 3.37, a calculation that does not include the octahedrally coordinated metal, since it is based on the ^{29}Si intensities. More drastic treatment, heating in steam at 700°C, produces even greater spectral changes and an Si/Al ratio of 6.89 (c). Repetition of this procedure followed by leaching with nitric acid (d) removes most of the aluminium to give Si/Al = 50 and a single ^{29}Si line, which indicates good crystallinity as is required if the Al vacancies are filled. The octahedral aluminium resonance becomes very narrow on leaching and represents remaining Al in the form of $Al(H_2O)_6^{3+}$ free to rotate in lattice cavities. These changes can also be achieved by treatment with $SiCl_4$ vapour or the aluminium can be put back into the structure using $AlCl_3$ vapour, both processes having been monitored by ^{27}Al MAS.

10.6   Deuterium, an integral-spin nucleus

The patterns obtained in the solid state with the nucleus ^{2}H, for which $I = 1$, are somewhat different from those described above. A nucleus with $I = 1$ has three energy levels and two degenerate transitions in the absence of any quadrupole coupling. The interaction of the nucleus with the magnetic field and the electric field gradient causes the three energy levels to be modified, so that there are two transition frequencies disposed symmetrically about the isotropic chemical shift value and with a frequency difference that is proportional to the quadrupole coupling constant and to the orientation of the bond to deuterium (and so the electric field gradient) relative to the magnetic field. In a powder sample all orientations exist and the resulting ^{2}H spectrum, shown in Fig. 10.19,

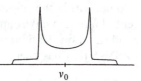

v_0

Fig. 10.19 The shape of the ^2H resonance in a solid powder sample, or Pake powder pattern.

has a particular shape and is known as a Pake powder pattern. The separation of the sharp edges of this spectrum is three-quarters of the value of the quadrupole coupling constant and usually lies in the range 120 to 150 kHz. If the moiety in which the deuterium lies is capable of rotation in the solid, then the width of the Pake pattern will be reduced and the extent of the reduction and the shape of the pattern will depend upon the details of the motion. The motion has to be fast relative to the value of the quadrupole coupling constant. These

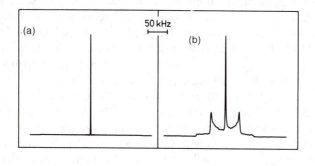

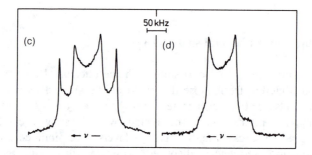

Fig. 10.20 The ^2H NMR spectra of deuteriobenzene (benzene-d_6) absorbed on graphite to a thickness of 10 molecular layers: (a) temperature 298 K; (b) at 170 K; (c) at 90 K and time between read pulses of 10 s; (d) at 90 K but with the time between pulses 0.2 s so that the broad component is saturated. (From Boddenberg and Grosse (1987) *Z. Naturforsch*, **42a**, 272–4, with permission.)

comments will be illustrated by reference to the ^2H spectra of deuteriobenzene absorbed on graphite to form a multilayer some ten molecules thick. Spectra were obtained at several temperatures and are shown in Fig. 10.20. At 298 K a singlet narrow signal is observed, and shows that the absorbed benzene is reorienting as if it were in the liquid phase. Indeed, this is found to be the case even if the amount of benzene absorbed is reduced until it forms a monolayer. At 170 K the spectrum is of mixed form, with a Pake pattern and a minor singlet. This latter is due to the absorbed monolayer, which is still undergoing fast two-dimensional motion, while the Pake spectrum arises from benzene crystallites that form further out from the surface. The splitting between the sharp edges of the Pake pattern is 70 kHz, and this is typical of the value found for benzene undergoing fast rotation around its hexad axis. The spectra obtained at 90 K were run using two different sets of conditions. In one, the pulse repetition rate was slow, so that all the deuterons were detected. In the other, the pulse repetition rate was much faster, so that the component with the longer relaxation time T_1 was saturated and effectively was removed from the spectrum. In the first case, two components are observed, both Pake patterns, and with splitting of 70 and 140 kHz. In the second case, the spectrum of the less mobile phase has disappeared and the 70 k Hz pattern remains. Note, though, that this is not the same as that seen at 170 k, since there is a distinct asymmetry in the base. This is due to the now solid absorbed monolayer rotating only around its hexad axis, and it follows that the broader pattern arises from the now static benzene crystallites.

This type of spectroscopy is also much used in the study of liquid-crystal phases, where the Pake-type patterns obtained from these partially ordered materials can give much information about the degree of order and the rates of motion, and how these change with the experimental conditions.

Bibliography

NMR is a subject of sufficient importance to have a considerable literature devoted entirely to various aspects of both its technology and its use. The following list represents only a fraction of the total available but is sufficient to give an entry into the field.

Four comprehensive works for general reference:

Abragam, A. (1961) *The Principles of Nuclear Magnetism*, Oxford University Press, Oxford.

Emsley, J. W., Feeney, J. and Sutcliffe, L. H. (1965) *High Resolution Nuclear Magnetic Resonance Spectroscopy* (2 vols), Pergamon Press, Oxford.

Jackman, C. M. and Sternhell, S. (1969) *Applications of Nuclear Magnetic Resonance Spectroscopy in Organic Chemistry*, 2nd edn (*International Series of Monographs in Organic Chemistry*, vol. 5), Pergamon Press, Oxford.

Pople, J. A., Schneider, W. G. and Bernstein, H. J. (1959) *High Resolution Nuclear Magnetic Resonance*, McGraw-Hill, New York.

Introductory texts, some with a variety of problems:

Abraham, R. J. and Loftus, P. (1978) *Proton and Carbon-13 NMR Spectroscopy – An Integrated Approach*, Heyden, London.

Ault, A. and Ault, M. R. (1980) *A Handy and Systematic Catalog of NMR Spectra – Instruction Through Examples*, University Science Books, California.

Gunther, H. (1973, trans. 1980), *NMR Spectroscopy – An Introduction*, Wiley, New York.

Harris, R. K. (1983), *Nuclear Magnetic Resonance Spectroscopy, A Physiochemical View*, Pitman, London.

Levy, G. C., Lichter, R. C. and Nelson, G. L. (1980) *Carbon-13 NMR Spectroscopy*, 2nd edn, Wiley, New York.

Lynden Bell, R. M. and Harris, R. K. (1969) *Nuclear Magnetic Resonance Spectroscopy*, Nelson, London.

The techniques of NMR, including the Fourier transform methods, are described in the following two books:

Martin, M. L., Delpuech, J.-J. and Martin, G. J. (1980) *Practical NMR Spectroscopy*, Heyden, London.

Mullen, K. and Pregosin, P. S. (1976) *FT NMR Techniques – A Practical Approach*, Academic Press, New York.

Two-dimensional NMR is discussed at various levels in:

Croasmun, W. R. and Carlson, R. M. K. (eds) (1987) *2-D NMR Spectroscopy, Applications for Chemists and Biochemists*, VCH, Munich.
Derome, A. E. (1987) *Modern NMR Techniques for Chemistry Research*, Pergamon, Oxford.
Kessler, H., *et al.* (1988) *Angew. Chem.*, **27**, 4 (a review).
Sanders, J. K. M. and Hunter, B. K. (1987) *Modern NMR Spectroscopy, A Guide for Chemists*, Oxford University Press, Oxford.

Many reviews are published that are devoted entirely to various aspects of NMR spectroscopy. While these would be expected to be outside the required reading of the students using this text, they nevertheless provide an entry into the recent literature. Two such are:

Mason, J. (ed.) (1987) *Multinuclear NMR*, Plenum Press, New York.
Popov, A. I. and Hallenga, K. (eds) (1991) *Modern NMR Techniques and Their Application in Chemistry*, Marcel Dekker, New York.

The student may also care to read the following few original short papers, which summarize the early and unexpected results that heralded the development of NMR as a subject useful to chemists:

Dickinson, W. C. (1950) *Phys. Rev.*, **77**, 736. Observed chemical shifts in fluorine compounds and noted the effect of exchange.
Proctor, W. G. and Yu, F. C. (1950) *Phys. Rev.*, **77**, 717. [14]N chemical shift between NH_4^+ and NO_3^-.
Arnold, J. T., Dharmatti, S. S., and Packard, M. E. (1951) *J. Chem. Phys.*, **19**, 507. First observation of chemical shifts in a single chemical compound.
Gutowsky, H. S. and McCall, D. W. (1951) *Phys. Rev.*, **82**, 748. An early observation of spin–spin coupling.
Gutowsky, H. S., McCall, D. W. and Slichter, C. P. (1951) *Phys Rev.*, **84**, 589. A theory of spin–spin coupling.

It should be remembered in reading the last two papers that the hertz separation of the ^{31}P doublet and ^{19}F doublet are the same. The gauss separation can be calculated from $\Delta B_0 = (J/v_0)B_0$ and is greater for ^{31}P since v_0 is smaller for the fixed field used.

Answers to questions

Chapter 1

1.1. 250 Hz and 1000 Hz; then 1000 Hz and 250 Hz. The frequencies corresponding to each signal are interchanged.

1.2. 10 000 Hz. This covers a spectral range of 5000 Hz adequately but not 25 000 Hz. In this latter case a pulse length of 40 μs could be tolerated but would give some intensity distortion at the extremes of the chemical shift range.

1.3. (a) None, (b) M_z only, and (c) M_{xy}.

Chapter 2

2.1. The $CHCl_2$ proton.

2.2. 7.3 ppm; 2555 Hz and 7.3 ppm.

2.3. The ^1H and ^2H chemical shifts are the same but the ^2H frequency separations are less, so that in the absence of spin–spin coupling (see Chapter 3) the resolution is worse.

Chapter 3

3.1. Chemical shifts are 1.48 and 3.56 ppm; $^3J(H–H) = 7$ Hz.

3.3. 15.6 Hz.

3.4. 1:6:15:20:15:6:1.

3.5. $J = 8.6$ Hz, shift $= 5.1$ Hz, 0.085 ppm. At 500 MHz the lines have shifts from TMS of 3244.3, 3235.7, 3200.9 and 3192.3 Hz with the intensity ratio strong/weak $= 1.49$. The exact chemical shift is 6.437 ppm.

3.6. Because in (a) the Hc resonance is broadened slightly by coupling to the methyl group and in (b) Hc is broadened slightly by coupling to the ring nitrogen.

3.7. The pattern is a doublet of quartets of triplets with coupling constants of 53, 7.3 and 4.7 Hz. No lines overlap.

Chapter 4

4.1. 22.5 s.

4.2. $R_{1DD}(H)/R_{1DD}(C) = 1.14$. Both H and C relax H whereas only H relaxes C.

4.3. Since the z coordinate is zero, then no terms can cancel and the EFG is proportional to $-3qr^{-3}$. For zero EFG, each term of the sum has to be zero, and if z is the coordinate of the displaced charge, then $3z^2 - r^2 = 0$ and $z = \pm r/\sqrt{3}$.

Chapter 5

5.1. Recovery time $= 0.693 T_1$.

5.2. Calculated linewidth is $0.7\,\text{Hz}$, $R_1 = 2.22\,\text{s}^{-1}$. Actual $T_2 = 3.54\,\text{ms}$, so $R_2 = 282.7\,\text{s}^{-1}$. The proportion due to scalar relaxation is then $R_2 - R_1$ or $280.5\,\text{s}^{-1}$.

5.3. $DW = 28\,\mu\text{s}$ (to nearest whole μs); 32K memory will be swept in 0.917 s, giving the required resolution. (In the final spectrum, 16 384 memory locations represent 18 000 Hz or 0.91 Hz per location.) The nuclei will not have relaxed fully after 0.917 s and the pulse should be shorter than 90°. The maximum permissible pulse length is $55\,\mu$s.

Index